Companion to
A Sand County Almanac

Companion to
A Sand County Almanac

Interpretive & Critical Essays

*

Edited by J. Baird Callicott

THE UNIVERSITY OF WISCONSIN PRESS

Published 1987

The University of Wisconsin Press
114 North Murray Street
Madison, Wisconsin 53715

The University of Wisconsin Press, Ltd.
1 Gower Street
London WC1E 6HA, England

Printed in the United States of America

For LC CIP information see the colophon

ISBN 0-299-11230-6

Frontispiece photograph courtesy of Robert A. McCabe

Contents

Contents

III. The Upshot

IV. The Impact

Appendix

Preface

Aldo Leopold's *A Sand County Almanac* has brought a generation to a groundswell change in environmental consciousness—
and conscience. For all its charm and apparent simplicity, *A
Sand County Almanac* is as incisive and complex as it is transformative. This volume brings together twelve interpretive and
critical essays by ten scholars, representing various academic dis-
ciplines. For us, *A Sand County Almanac* has been not only an
inspiration, but an object of careful study. I hope that these essays will begin the long overdue process of exploring its intricacies and sounding its depths. As with any work which so sweep-
ingly reconstructs our experience, *A Sand County Almanac* also
recasts our questions. Most of all, I hope that this volume will
press a little further into the philosophical *terra incognita* toward
which Leopold's legacy so alluringly draws us.

It was in the spring of 1971 that Robert Ramlow, formerly a student in my course History of Ancient Greek Philosophy, suggested I read *A Sand County Almanac* by Aldo Leopold. He
offered to lend me his copy. I was preparing to teach a new
course called "Environmental Ethics" at Wisconsin State University–Stevens Point, and had no syllabus and no textbooks. (In
retrospect, I should not have been so surprised to discover that
there was no established way of proceeding or model curriculum
to follow. As it now appears, mine was the first philosophy
course of its kind ever offered.)

I was an expatriate Southerner, fresh from the pitched battles
of the Civil Rights struggle in Memphis, Tennessee, and I had
heard of neither the book nor its author. I wondered what good

an "almanac" written by someone who for all I knew might have been a farmer, could possibly do *me,* a philosopher steeped in the Classics, as I desperately tried to build a strange new ethics course from the ground up. Although he found it difficult to say just what sort of book it was, Bob assured me that I would find reading *A Sand County Almanac* very worthwhile. I trusted his judgment; he was a senior majoring in Resource Management for Environmental Quality and surely knew more about these things than I. Besides, I was staring into the yawning emptiness of a sixteen-week semester, commencing in just three short months, so I could ill afford to leave any stone unturned in my search for suitable literature.

When I joined the faculty of a state college in the central sand country of Wisconsin as instructor of philosophy in 1969, the environment was under wholesale assault from every direction with no surcease in sight. Civil Rights was a cause already won in the republic of ideas and in the courts (if not on Main Street in Memphis). The war in Vietnam would eventually end—one way or another, sooner or later. On the other hand, the "environmental crisis" (as it was then called) appeared to be ubiquitous in scope, gargantuan in scale, and protracted in duration. Its dimensions—and character—appeared to me to shake Western cultural values and beliefs to the foundations.

To blame greedy capitalists, big business, complicitous politicians, and bureaucratic insouciance for the worsening environmental malaise, to me seemed too facile—as well as too fashionable. The causes of the environmental crisis were diffuse, insidious, and synergistic. It was rooted in our whole way of doing things—modern technology—and our basic values—humanism and the *summum bonum* of human happiness, defined as pleasure and measured by money and material accumulation.

I was a concerned citizen, but I was also, more particularly, a challenged philosopher. I asked myself how, as a philosopher, I could contribute to a rethinking of human nature and a reconstruction of human values to help bring them into line with the relatively new ideas about the nature of the environment emerging from ecology and the New Physics. Then, perhaps, we could gradually begin to adjust ourselves to the ecological exigencies

Preface

and environmental limitations which were beginning to make themselves palpably plain. Anything short of a philosophical overhaul of prevailing attitudes and values toward nature seemed to me then, as now, to treat the symptoms of the maladaptation of global civilization to the planet—not the disease itself.

I found other more-or-less suitable texts for my course in environmental ethics during the summer of 1971, but, while the other texts shuffled in and out, *A Sand County Almanac* remains at the core of my curriculum sixteen years later. By a sort of spontaneous generation, other courses in environmental ethics and related subjects sprang up at colleges and universities here and there. The focus of my own philosophical efforts shifted gradually from pedagogy to helping build a literature in this new "field." Aldo Leopold's *A Sand County Almanac,* however, continued to inspire and inform my thought. Leopold always seemed to have the right intuitions, to be thinking in the right directions. But his ideas were so compactly expressed that their full implications were not evident. I tried to unpack and air out Leopold's fertile thought in the rapidly expanding debate (most of which was carried on in the journal *Environmental Ethics* founded by Eugene C. Hargrove) that surrounded philosophical issues in environmental ethics: anthropocentric vs. nonanthropocentric value theory, moral atomism vs. moral holism, the relation of ecological fact to environmental value, and so on.

This book provides me with an opportunity to share with a larger audience one of my previously published articles on Leopold's monumental effect on modern environmental awareness and an occasion more directly to write a sustained and thorough philosophical analysis of his seminal contribution to what has now become a well-established subdiscipline of academic philosophy. I am honored to be able to place my essays alongside those of the distinguished authors included here. I have tried to fit all the essays together to form a coherent book—a chorus of distinct voices blended into a harmonious whole. As such, they are intended to enrich and extend a reading of Leopold's classic.

I thank all the contributors to this volume for their diligence, patience, and cooperation throughout the assembly process.

Preface

Among them I especially thank Roderick Nash for his unfailing enthusiasm, encouragement, and support, Dennis Ribbens and Curt Meine for their ready advice and consultation on the structure of the whole and on matters of factual detail, and Peter Fritzell for his patient and thorough editorial assistance. I thank Nina Leopold Bradley for her help and warm encouragement from first to last and for lending me the photographs of Aldo Leopold and the shack which appear on the cover and frontispiece of this book. I thank Alice Van Deburg, acquisitions manager at the University of Wisconsin Press, who immediately saw the importance of a volume such as this and energetically facilitated its completion. I thank Frances Moore Lappé for her thoughtful editorial suggestions on my essay, "The Land Aesthetic," and Elizabeth Steinberg, chief editor of the University of Wisconsin Press, for her expert editorial advice on the volume as a whole and on my contributions to it in particular. I thank my father, Burton Callicott, for contributing the illustrations appearing on the part-title pages. Finally, I thank Carolee Cote for her patient word-processing of all my contributions to this book as well as those of several other contributors.

J. Baird Callicott
August 15, 1986
Oakland, California

Companion to
A Sand County Almanac

SCALE

| 0 | 25 | 50 | 75 | Mi. |

| 0 | 25 | 50 | 75 | 100 | Km. |

The sand county area of central Wisconsin.

Introduction

J. BAIRD CALLICOTT

Aldo Leopold died, suddenly, in the spring of 1948, shortly after his slender volume of essays, then titled *Great Possessions,* had at last been accepted for publication by Oxford University Press. Luna B. Leopold, his second eldest son, saw the manuscript through the process of editing and production. The book was published in 1949 as *A Sand County Almanac.* Luna Leopold edited a second book of Aldo Leopold's essays and journal entries entitled *Round River,* which Oxford University Press published in 1953. Some of the essays from *Round River* ("A Man's Leisure Time," "Country," "Natural History—The Forgotten Science," "The Deer Swath," "Deadening," "Conservation," "The Round River—A Parable," and "Goose Music") were re-edited and added to a second edition of *A Sand County Almanac* published by Oxford in 1966. Since 1970 Ballantine Books has distributed an inexpensive paperback of this enlarged edition. In its various editions, *A Sand County Almanac* has sold more than a million copies.

In the present volume, the titles *A Sand County Almanac* or *Sand County* or the *Almanac* (unless otherwise indicated) all refer to the original edition of 1949 (or its subsequent cloth and paper facsimiles). Quotations from *A Sand County Almanac* are identified by page numbers in parentheses. Quotations or references to essays which originally appeared in *Round River* are cited, accordingly, in footnotes.

A Sand County Almanac—a bestseller, the environmentalist's bible, and a twentieth-century landmark in a genre of Ameri-

3

can letters pioneered by Thoreau and advanced by Marsh and Muir—has received, until the appearance of this volume, surprisingly little systematic interpretation or critical discussion. Roderick Nash devoted a chapter to Aldo Leopold in his celebrated *Wilderness and the American Mind* (now in its third revised edition) and Susan Flader has written a book-length monograph, *Thinking Like a Mountain: Aldo Leopold and the Evolution of an Ecological Attitude toward Deer, Wolves, and Forests,* and several shorter articles about Leopold's life and thought. But although *A Sand County Almanac* is ubiquitously quoted and the land ethic frequently invoked, essays devoted to a patient study of the book as a whole or to its capstone essay, "The Land Ethic," may literally be counted on the fingers of one hand. Most of them are included in this volume.

Writing in the mid-seventies, Peter Fritzell complained about this circumstance:

> Despite [*Sand County's*] popularity and reasonable longevity . . . no scientist has considered it much more than pleasant or moving material to be read at leisure. No philosopher has approached its ethics. No literary critic has suggested it might be fruitfully examined with a systematic eye to style, metaphor, conception, or narrative.[1]

And Dennis Ribbens, writing in the early eighties, echoed Fritzell's lament: "Few books have had as much influence on America's growing ecological awareness. . . . But for all that the book has been given little critical . . . attention."[2]

The present volume is intended as a remedy.

A Sand County Almanac moves progressively from the personal to the universal, from the experiential to the intellectual, from the concrete to the abstract. So does this, the companion volume. The first section provides an account of the author's life, his geographical and cultural milieu, and finally his place in a history of ideas. The second section focuses on the *Almanac* as a whole—its genesis, evolution, elements, and movements. The third section "wrestle[s] with the philosophical questions of Part III" of the *Almanac* (viii). The fourth is a denouement; it considers the impact of *A Sand County Almanac* on public resource management, on American environmental consciousness more gener-

Introduction

ally, and as crystalized in a state-of-the-art example of contemporary environmental philosophy.

Each section, moreover, is organized so as to reflect the flow of the whole. The first essay of each section is so ordered because, compared with the others in its group, it is more concrete, personal, or experiential. The essays that follow are more challenging and more specialized. Hence, there is something in each section of this book which every reader of *A Sand County Almanac* will find accessible, and, at the same time, there is material that will be of use to scholars.

This is a book about a book, *A Sand County Almanac,* not about a man, Aldo Leopold. Leopold was much more than the author of *A Sand County Almanac.* He was also author of *Game Management* and numerous technical papers and reports, founder of the discipline of wildlife management, influential advocate of wilderness preservation and cofounder of the Wilderness Society, eminent conservationist, professional forester, college professor, homeowner, steward of 120 acres of Wisconsin River floodplain, son, brother, husband, father, friend, hunter, bowmaker, ... *A Sand County Almanac,* on the other hand, is much more than Aldo Leopold; it has achieved the independent life of all great art and philosophy. But to better understand a book, it is more than a little helpful to become acquainted, at the outset, with its author. Thus the first section of this collection is called "The Author."

Here the essays of Curt Meine, Susan Flader, and Roderick Nash locate the author in progressively wider contexts. Curt Meine introduces us to the boy, the prep school and college student, the young forest ranger, and the mature man on the point of a midlife career change. His straightforward biographical sketch brings us up to the investment of the man in the land.

In "Aldo Leopold's Sand Country," Susan Flader introduces us to the land—to Wisconsin's sand counties celebrated by Aldo Leopold. She provides an intimate, authoritative account of Leopold's shack experiences with his family and with his fellow tenants of shore, woods, marsh, and field. Readers who do not know the sand country of central Wisconsin firsthand will find in

Flader's essay a vivid portrait of the larger shack environs and a systematic account of its geocultural history.

Roderick Nash draws the circumference of the author's milieu wider still. Leopold's evolutionary ecological "land ethic" did not occur in an intellectual vacuum; nor was it without precedent. Nash, in "Aldo Leopold's Intellectual Heritage" puts the idea in its place—in the history of ideas, as Flader places the person in the geocultural history of the land. Nash argues, in sum and substance, that Leopold's land ethic is less original than it is touted to be. Henry David Thoreau and John Muir certainly had argued for a morally charged human relationship with nature. But they had done so in essentially religious terms. Leopold's originality lies in his having expressed their moral proposition exclusively in scientific terms—as Wallace Stegner points out in his contribution to this volume.

Charles Darwin, three-quarters of a century earlier than Leopold, had observed the evolutionary-historical "ethical sequence" and forecast the extension of "sympathy beyond the confines of man."[3] Darwin's vision of an ethic extending beyond the confines of man did, in fact, rest expressly on scientific foundations. And, as Nash makes us aware, at the turn of the century, in addition to the Englishman Henry Salt, two Americans, Edward Payson Evans and J. Howard Moore, had also speculated, in the light of Darwin's theory, about extending ethics beyond the sphere of human relationships.

But, in my view, they all, without exception, broke a trail which diverges widely from the one taken by Leopold, though both trails radiate, to be sure, from a common hub of ideas. Darwin goes on to say immediately just what he has in mind by "sympathy beyond the confines of man": "that is, humanity to the lower animals." Darwin, and certainly Salt, Moore, and Evans, all anticipate not so much Leopold's land ethic as animal liberation and animal rights.

The land ethic is informed not only by evolution—which discloses, among other pertinent things, a bond of kinship between mankind and the "higher" "lower animals"—but by ecology, a science not yet even identified as such when Darwin wrote and not widely known or appreciated at the turn of the century. To-

Introduction

day, the sentimental and "atomistic" animal liberation/rights ethics and the ecologically informed "holistic" land ethic are recognized by partisans of both to be not only in theory divergent, but in practice contradictory.[4]

Nash complains that all of us—Leopold and "Leopold scholars" alike—are less aware of the historical antecedents of the land ethic than we ought to be. The philosophically sensitive reader, on the other hand, may wish that Professor Nash had given more attention to the very great theoretical differences among the sentiency-based humane ethics of Darwin, Salt and Evans, the neo-Kantian "reverence-for-life" ethic of Albert Schweitzer (whom he also mentions as an antecedent), and the evolutionary-ecological land ethic of Aldo Leopold. Leopold's historical tools—"saw, wedge, and axe"—albeit simple and rustic, are specialized and thus precise (16). Nash works in a well-surveyed woodlot, but using the modern power tools of contemporary historians he has cut various softwoods along with good oak and stacked them in the same cord.

The essays in the next section, "The Book," make it abundantly clear, in various ways, that *A Sand County Almanac*—at first glance a miscellaneous hodgepodge—is wonderfully unified and tightly organized.

Dennis Ribbens' fascinating study of Leopold's correspondence with his prospective publisher and collegial critics reveals that unity and thematic structure were matters of disagreement and misunderstanding between Leopold and the editors at Alfred A. Knopf who solicited a "nature book" from him in the early 1940s. Leopold consciously strove to fashion a thematically unified book from autonomous vignettes. The book itself is a kind of literary ecosystem; it consists of discernible, articulate parts, competing and cooperating with one another, all interrelated to form a single, emergent, but extraordinarily coherent whole.

What makes the difference between a good book and a great book? How does a great book become a classic and a classic, for some, holy writ? I am sure a little luck and good timing don't hurt. Neither does style. The appeal and the memorability of

Sand County may be attributed in large measure to its craft. John Tallmadge, in this section's next essay, undertakes a comparative and analytic literary examination of *A Sand County Almanac*. *A Sand County Almanac* is a contribution to a genre of belles lettres usually called "nature literature." Tallmadge treats it, rather, as a specimen of "natural history" and, comparing *Sand County* to other classics of that genre, exposes the principal ingredients of its compact style.

Finally, in this section, Peter A. Fritzell submits the book as a whole to a sympathetic deconstructive thematic critique. Fritzell claims that beneath the manifest sustained argument of *A Sand County Almanac* there runs a countercurrent, a riptide, which generates, in part, the artistry of the book—in particular, its frequent irony and something, at the philosophical level, very like suspense or tension in a good mystery story. Fritzell does not deliberately employ a deconstructionist method—a recent rage in literary criticism—nor is he happy with my association of his essay with this fashionable new school of textual analysis. However that matter may be, Fritzell's essay is a marvel of subtlety and sophistication and in my opinion the most insightful study of *A Sand County Almanac,* as a whole, ever made. (I address the conceptual paradox or even contradiction which Fritzell finds "on virtually every page" of the book in my essay, "The Conceptual Foundations of the Land Ethic" in section 3.)

The third section of this volume focuses on "The Upshot" of *A Sand County Almanac* and takes its title from Part III of that book. With due regard to my personal bias as a philosopher, I think it is fair to say that Leopold's grappling with "the philosophical questions of Part III" is responsible in the final analysis for the greatness of *A Sand County Almanac* and certainly for its contemporary apotheosis. His book would not be the cultural institution it has become had Aldo Leopold not been a consummate prose stylist as well as a renowned naturalist. But neither would it have been more than a fine nature book had its author not also been a philosopher. Science, "poetry," and philosophy all coalesce in the climactic essays and especially in "The Land Ethic" of Part III of *A Sand County Almanac*.

Introduction

"The Upshot" of *Sand County* begins with the "Conservation Esthetic." Natural (as opposed to artistic or, more technically, artifactual) aesthetics is a persistent concern of the author of *A Sand County Almanac*. The "esthetic harvest" which land "is capable, under science, of contributing to culture" is introduced in the foreword as one of the three principal themes of the book, and the "beauty" (as well as integrity and stability) of the biotic community is included among its most valued qualities in the famous summary moral maxim of the last section of *Sand County's* last essay, "The Land Ethic" (viii). Leopold's contribution, however, to natural aesthetics has been scarcely acknowledged or appreciated, in part because the "Conservation Esthetic" presents a much less sustained argument than "The Land Ethic." How can land, *under science*—a most important qualification—yield an aesthetic harvest? In "The Land Aesthetic," I sift through the other remarks on natural aesthetics, natural beauty, taste in natural objects, and so on, that are scattered throughout *A Sand County Almanac* (and *Round River*), in an effort to indicate how it may.

"The Building of 'The Land Ethic,'" by Curt Meine is a genetic study. As Meine points out, "The Land Ethic" (which, he correctly claims, is "Leopold's most important essay") was not written all at once, nor was the previously published "The Conservation Ethic"—as one might suppose from a casual comparison of both—simply taken out of mothballs and refitted for a "collection" of essays old and new. Rather, "The Land Ethic" was compiled from parts of three previously published pieces ("The Conservation Ethic" among them), which were cut down, very carefully reworked, and synthesized with newly written bridging and concluding material. Meine's essay does for "The Land Ethic" something of what Ribbens' does for the *Almanac* as a whole. And it makes equally fascinating reading for the affectionate lay fan as well as for the serious scholar.

My next piece, "The Conceptual Foundations of the Land Ethic," is also analytic; but it is an anatomical rather than genetic analysis. However Leopold may have pulled it together, "The Land Ethic" is the distilled essence of his ripe thought. Its expression is extraordinarily condensed and deceptively simple.

Like the proverbial iceberg, only the tip of the land ethic shows above the verbal surface of Leopold's essay. My analysis undertakes to sound the body of ideas lying beneath.

In "The Conceptual Foundations of the Land Ethic" I am as little concerned with material historical influences as with how Leopold synthesized earlier papers to produce his nomothetic essay. It is certain that Leopold read Darwin, whose influence on his moral philosophy is most pronounced. And we know that he thought of himself as an ecologist (as indeed he was) and that he was a personal friend of Charles Elton, who, in the late twenties, articulated, more fully than anyone before, the ecological "community concept" that is so central to the land ethic. Whether Leopold read David Hume, Adam Smith, and the other philosophers whose ideas in one form or another I find in "The Land Ethic" is an open question. The ideas of great philosophers become part of our intellectual vocabulary; and broadly educated people, like Leopold, pick them up, tune them in, in any number of indirect ways.

The fourth section of this book, "The Impact," is intended to provide some sense of the lasting cultural effect of *A Sand County Almanac*. It begins with a very personal narrative by Edwin P. Pister, "A Pilgrim's Progress from Group A to Group B."

Pister is a state fish- and game-agency employee, a conservationist toiling in the trenches. As he recounts in his essay, he began his career taking for granted the Pinchot utilitarian philosophy of "resource management" for human happiness ("preference satisfaction" in the jargon of contemporary economics) which prevails in his profession. One of Pister's mentors was A. Starker Leopold, whose teachings and whose father's *A Sand County Almanac* gradually converted him to a wider outlook on his professional mandate. His autobiographical essay concretely illustrates the application of the land ethic to contemporary conservation practice and policy.

Edwin Pister spearheaded both the heroic fieldwork and equally heroic political efforts, ultimately reaching the United States Supreme Court, which rescued the Devil's Hole pupfish (*Cyprinodon diabolis*) from extinction.[5] The plight of the desert

Introduction

pupfish and its eleventh-hour stay of execution has become an environmental *cause célèbre*—a symbol of the conservation of nongame nonresources and the responsibility of agency employees to guard the "biotic right" of species, however uneconomic, to continuation.

Wallace Stegner's "The Legacy of Aldo Leopold" provides a panoramic perspective on the state of the American environment and the American environmental state of mind nearly four decades after *A Sand County Almanac*. Stegner writes a personalized summary of Leopold's philosophy, emphasizing that it is contrary to institutionalized American values. His essay gives an overview of the signs of Leopold's effective consciousness raising, indicated by national environmental legislation and the changes in the orientation and growth of older environmental organizations—e.g., the National Audubon Society, the Isaac Walton League, and the Sierra Club—and the emergence of new ones—e.g., the Nature Conservancy and the Environmental Defense Fund. Nevertheless, Stegner feels that Leopold's message has reached only a literate, dedicated minority. Most Americans—whether they occupy the penthouse suites of America's corporate headquarters or the back booths in the corner bars of America's industrial hinterlands—remain fixated on the Almighty Dollar, poised to seize the Main Chance.

In "Duties to Ecosystems," Holmes Rolston, III, engages Aldo Leopold's land ethic in a philosophical dialectic. Rolston poses and attempts to answer the general ecophilosophical question raised by "The Land Ethic," after nearly forty years of both philosophical and ecological development: Can we intelligibly entertain duties to nonindividuals—as the land ethic proposes—more particularly to ecosystems? (I should note here that while Roderick Nash writes freely of the "rights of nature," the idea that nature as a whole has rights is, from a philosophical point of view, out of the question. Rights are claims expressed by, or on behalf of, moral *patients*. "Nature," per se, is too ill-defined an "entity" to be even considered as a candidate for rights. Duties, on the other hand, belong to us as moral *agents* and ecosystems are specifiable, albeit complex, structures.)

Holmes Rolston, III, is uniquely qualified to explore the vanguard philosophical problems of duties to ecosystems. He is not only an able philosopher, among the founders of environmental ethics as an academic philosophical discipline, he is intimately conversant with the relevant ecological literature, as reading his essay will confirm from the very first page. What may not be so evident from a reading of his essay is that he is also an accomplished and highly sensitive field naturalist. Hence his understanding of the nature of ecosystems—crucial to the question of our duties regarding them—is firmly grounded in both science and experience.

The foreword to *A Sand County Almanac* (dated March 1948) provides, in my judgment, a wonderfully understated lyrical account of Leopold's own understanding of the principal burden and argument of his book. Among the Leopold Papers in the University of Wisconsin–Madison Archives, however, there has reposed another foreword (until now never published) dated July 31, 1947. It has been briefly quoted in several published works, most notably by Roderick Nash and Susan Flader.[6] For this reason, as well as for its more general inherent interest, it seemed fitting to me to include it as an appendix to this volume. Moreover, Leopold may in fact have wished that it had been published in revised form as an appendix to *A Sand County Almanac*. His untimely death, it seems, thwarted those intentions. Dennis Ribbens called it to my attention and agreed to introduce it for publication in this volume. Nina Leopold Bradley, in consultation with other members of the Leopold family, kindly granted me permission to print it.

The 1947 foreword is a more conventional introduction than its replacement. It is more discursive, and, in Leopold's characteristic casual style, it names and ties together most of the essays which it was designed to precede. More important, it also ties many of the essays to the author's life experiences. Hence, it not only reveals the author's own sense of the organic connection of the essays to one another, it unifies them with the odyssey of the author's life.

Introduction

Notes

1. Peter Fritzell, "Aldo Leopold's *A Sand County Almanac* and the Conflicts of Ecological Conscience," *Transactions of the Wisconsin Academy of Sciences, Arts, and Letters* 64 (1976): 22.

2. Dennis Ribbens, "The Making of *A Sand County Almanac,*" *Wisconsin Academy Review* 28:4 (1982).

3. Charles R. Darwin, *The Descent of Man and Selection in Relation to Sex* (New York: J. A. Hill and Company, 1904), 124.

4. See Tom Regan, "Ethical Vegetarianism and Commercial Animal Farming," in *Contemporary Moral Problems,* ed. James E. White (St. Paul, Minn.: West Publishing Co., 1985), 279–93, and J. Baird Callicott, "Animal Liberation: A Triangular Affair," *Environmental Ethics* 2 (1980): 311–38, for examples of both, respectively. Regan comments that "environmental fascism [his rhetorically charged rubric for the land ethic] and *any* form of rights theory [including animal rights] are like oil and water, they don't mix" (p. 292, emphasis in original).

5. For an account see Edwin P. Pister, "Desert Pupfishes: Reflections on Reality, Desirability and Conscience," *Fisheries: A Bulletin of the American Fisheries Society* 10 (November/December 1985): 10–15.

6. Roderick Nash, *Wilderness and the American Mind* (New Haven: Yale University Press, 1967), 192; Susan Flader, *Thinking Like a Mountain: Aldo Leopold and the Evolution of an Ecological Attitude toward Deer, Wolves, and Forests* (Columbia: University of Missouri Press, 1974), 102.

I * The Author

1 ✶ Aldo Leopold's Early Years

CURT MEINE

Ordinary eyes would have found little to commend in the wasted
acres that Aldo Leopold and his friend Ed Ochsner first looked
over on that chilly day in January 1935. The land was all but barren. On one side of the road, young popples fringed a frozen
marsh. On the other side, in what once were a farmer's fields,
corn stalks and sandburs stood stiffly above the crust of snow.
A line of gaunt elms led toward the remains of a burned-down
farmhouse. Beyond, the Wisconsin River lay ice-locked in its
winter channel. Yet, in Leopold's eyes, no land was without
promise. "In country, as in people," he once wrote, "a plain exterior often conceals hidden riches, to perceive which requires
much living in and with."[1] So with the abandoned Wisconsin
farmstead: on that poorest of land, in that deepest of winter, in
those hardest of times, he saw possibilities.

Aldo Leopold spent a lifetime sharpening his senses, recording his impressions, turning over within his mind the world he
saw without. All of us do the same, but few of us have the desire,
opportunity, or discipline to do it so thoroughly. Moreover, perception for Leopold was no mere aesthetic exercise, but an active, creative process. He took his findings and applied them, in
word and deed, back to the world from which they came. He

The research on which this essay is based was undertaken by the author while at the Institute for Environmental Studies at the University of
Wisconsin–Madison. His biography of Leopold, *Aldo Leopold: His Life
and Work* will be published in 1988 by The University of Wisconsin Press.

maintained the connection between the visionary and the practical that gives vitality to both, and provided that connection, as well, for an untold number of colleagues, students, listeners, and readers.

Had Leopold grown up in other surroundings, or with other interests, he might have employed his perception as a poet, or a doctor, or an inventor, or an architect. Instead, he grew up in close contact with the outdoors, and his parents encouraged his love for "things natural, wild, and free" (vii). He became, finally, a naturalist. We label him as such, but we also need to qualify the label. We tend to think of naturalists as simple souls who content themselves in the call and response of the wild, who, blithely removed from human society, extol the natural wonders of their Selbornes, Waldens, Yosemites, Wisconsins. Aldo Leopold was not a simple man, however simple were his convictions. He lived only sixty-one years, but his birth and death bracketed a period of profound change on the American landscape and in our civilization. His words endure not because they ignore those changes, but because they account for them from a unique perspective. They speak to our need to know the place we live in and to understand the changes that bind the past to the future. They help us to appreciate what it is to be alive on this magnificent, improbable planet.

Rand Aldo Leopold—the first name was dropped early on—was born in the Mississippi River town of Burlington, Iowa, on January 11, 1887. He was the oldest of Carl and Clara Leopold's four children, three boys and a girl. The family lived in a stately Victorian home atop Prospect Hill, one of several limestone bluffs surrounding Burlington and affording an impressive view of the Mississippi Valley to the east.

Aldo's mother was a petite, pretty bundle of energy. Her father was Charles Starker, a native of Stuttgart who came to America in 1848. A man of impressive talents, Starker prospered in Burlington as an architect, landscaper, engineer, banker, and businessman. He gave his daughter a private education, a love of natural beauty, and a thorough immersion in German romanticism. Her mother, Marie, taught her the social graces that a young lady required and the homemaking skills—Clara was a particularly fine cook—that she would later employ as the Leopold family matriarch. Clara was intelligent, extroverted, mischievous, and no fragile flower; a talented ice skater, she enjoyed sports and loved to join her father in his continual gardening and groundskeeping activities on Prospect Hill. Despite hearing problems, she played piano well, and her greatest thrill, the highlight of her year, was an annual trip to Chicago to attend the grand opera. This abiding love of opera was the legacy of several trips to Germany that she had made with her parents. Steeped by her father in the works of Schiller and Goethe, Clara was an unapologetic romantic. In a later age, her sharp mind and spirited personality might have taken her far from Burlington. In the 1880s, though, choices were few, and she remained at home on Prospect Hill, where eventually she fell in love with her itinerant cousin, Carl Leopold.[2]

Carl was a solid, personable young man, lanky, unaffected, conscientious—a prairie gentleman. He was the youngest of eight children born to Charles J. J. Leopold and his wife Thusneld, who was a sister of Marie Starker. Carl Leopold was born in Burlington, but spent his childhood years on the frontier in rough-and-tumble Liberty, Missouri. He and his pals were witness to the West's first bank holdup, pulled off by the soon-to-be-legendary James Gang. Western Missouri had yet to be subdued by the law, the plow, or the fence, and in its ample wilds Carl developed the love of outdoor life that he would pass on to his own children. He learned on his own how to hunt and trekked the environs of Liberty—upland prairies and oak savannahs and wooded valleys and Missouri River bottoms—with the eagerness and curiosity of an instinctive woodsman. When the time came for Carl to make his way in the world, he went east to Burlington to attend business school, and he eventually became a traveling salesman for a local hardware concern, riding by buckboard out to the islands of settlement in the prairie seas of Kansas and Nebraska. In between his sales trips west, he boarded with the Starkers in Burlington, and fell in love with his cousin Clara. They married in February 1886. Shortly thereafter, Charles Starker arranged for his son-in-law to come in off the road and become the managing partner in a company that made office furniture. Carl's enthusiasm, integrity, and business sense eventually made the Leopold Desk Company one of Burlington's most successful manufacturers. The credo of Carl's business was as forthright as his character: "Built on Honor to Endure."

In the early years of Aldo's life, the Leopolds lived with "Oma" and "Opa" Starker in the big house on Prospect Hill, but with the arrival of Aldo's sister Marie and brother Carl, Jr., Charles Starker decided it was time to build a house for his daughter and son-in-law. A fourth child, Frederic, was born soon after the Leopolds moved into the new house, which was only a backyard walk away from the Starker home.

The atmosphere in the Starker-Leopold compound was rich and Germanic, cultivated but informal. The beauty of the setting and the complement of family qualities go a long way toward explaining the interests, creative tensions, and well-defined character traits that Aldo carried into his adulthood. Here was beauty and business, the classics and the frontier, music and hunting, honor and easy humor, idealism and practicality, civic duty and rugged individualism, culture and nature—all carried on before the ever-flowing drama of the Mississippi. Grandfather Charles Starker set the tone. He had, in the words of a tribute paid to him by business colleagues, "a wholesome dislike for all pretense and hated hypocrisy in any form. His religion was too broad to be bounded by any creed, but he stood for all that was honorable and manly, just and noble."[3] Carl Leopold shared his father-in-law's independence from orthodoxy, and distrusted preachers of whatever creed. He had other ways of expressing his abiding faith.

Aldo's was a boyhood of dogs, yardwork, carriage rides into the countryside, Oma's baking, Opa's drawings, Clara's piano, Carl's outdoor excursions. The family was close, but it was plain from the outset that Aldo was Clara's favorite, a fact that Aldo took advantage of on occasion but learned to downplay. The other children simply learned to accept it. Aldo was a bright boy, self-assured but shy among strangers, a superb student. He excelled at Prospect Hill School, which was particularly strong in its teaching of history, English, and the classics.

Aldo Leopold's Early Years

It soon became clear, however, that Aldo's main interests lay outside the classroom. Carl Leopold gave to all his children a more than passing concern for the natural world. On hunts, picnics, fishing trips, or simple walks, he was the all-purpose guide, explainer of mysteries, and instiller of ethics and sportsmanship. In his lifetime, Carl had seen the abundance of the aboriginal prairie, the feverish rise in some game populations that first followed settlement, and the sudden decimation wrought by the market gunners. Long before conservation was a movement, he practiced voluntary restraint in the field and taught his children how to see the woods, how to perceive its hidden tales. As a hunter, he chose to limit his bag, stopped hunting certain species of ducks, quit spring hunting of waterfowl altogether, and always pursued crippled game until it was found or irretrievably lost. As a woodsman, he took the family on regular outings to show them the indwelling life of a hollow log, the mink-raided muskrat den, the droppings of a raccoon. "We did not need to kill game to have an exciting afternoon in the field," Frederic recalled.[4]

In the early 1890s, Charles Starker bought a membership in a summer resort in the Les Cheneaux Islands of northern Lake Huron near Michigan's Mackinac Island. Ostensibly purchased to provide Clara relief from her debilitating hay fever, the cottage became an annual escape to the near-wilderness for the entire clan and gave Aldo his earliest exposure to wild country. For six weeks each summer, he explored the northwoods, fished its wild waters, absorbed its untamed flavor, and dreamed of the open Canadian expanse, the blank spot on his childhood map, that lay beyond the Huron headlands to the north. For years, Aldo would dream of taking a canoe trip north into the Canadian backcountry, to meet a north-flowing stream that would carry him to Hudson's Bay.

Like many a naturalist, Leopold was drawn as a youth to study birds. As an eleven year old, he recorded one of his earliest systematic observations: "I like to study birds. I like the wren best of all birds. We had thirteen nests of wrens in our yard last summer. We hatched one hundred and twenty young wrens last summer."[5] He went on to list thirty-eight other species that he

CURT MEINE

was able to identify. In his early teens he began to devote himself more and more to this study, and he fancied himself an amateur ornithologist. He soon knew over two hundred and fifty species.

At the same time, he was beginning to hunt. He had long accompanied his father on hunts unarmed, absorbing Carl's methods and sportsmanship. Their main grounds were across the Mississippi in the bottomland swamps on the Illinois side. Carl belonged to two hunting clubs, Lone Tree and Crystal Lake, that owned the lands, and it was here that Carl took Aldo in the early mornings for his first hunts with gun in hand. In the summer, these backwater sloughs were hot and mosquito ridden, but fair country for Aldo and his closest friend, Edwin Hunger, to explore by rowboat and hip waders.

Aldo's interest in conservation also began to emerge in his early teens. Teddy Roosevelt was in the White House. Carl Leopold had always had the instincts of a conservationist, and in a business so closely associated with wood he was well aware of the decimation of the nation's forests, and of its northern pinelands in particular. The destruction of the pineries was plain to see during the family's annual summer trips to Michigan. Aldo's reading reinforced his inclinations. Besides his unusually broad education at school, Aldo read the literary accounts of Roosevelt, Ernest Thompson Seton, Stewart Edward White, Jack London, and the outdoor journals of the day. It was a time of growing public zeal for the wild, and this led to Aldo's decision in his mid-teens to pursue a career in forestry. Carl had hoped that Aldo would follow him in the family business, but he could hardly object to a decision that he himself had largely inspired.

In the meantime, Clara, eager to advance her favorite, was pushing for Aldo to receive, just as she had, an education out East. Aldo was quite content at Burlington High, which was an excellent school (and which, because it was overcrowded, did not require a full day of attendance—giving Aldo extra time to hunt). In the end, Clara had her way. Aldo was to head to New Jersey to the Lawrenceville School in January 1904. As if to clinch Aldo's decision to pursue a career in conservation, the family vacationed in Wyoming and Colorado the fall before.

Aldo Leopold's Early Years

Aldo and his father journeyed on their own to Yellowstone Park, and enjoyed an epic big-game hunt in the mountains east of the park.[6] It would not be Aldo's last taste of the West.

For all that Leopold would eventually write about his vast experience with the natural world, he actually wrote sparingly about his own background, his boyhood, his influences, his inner life. He was, in this sense, deeply reflective, but seldom autobiographical. In *A Sand County Almanac,* most references to his early life are only implied. The essay "Red Legs Kicking" is the only one in the entire collection in which Leopold recalls his youth at length. "My earliest impressions of wildlife and its pursuit," he wrote, "retain a vivid sharpness of form, color, and atmosphere that half a century of professional wildlife experience has failed to obliterate or to improve upon" (120). He describes the "unspeakable delight" he felt when, armed with his single-barreled shotgun, he brought down his first duck at the Crystal Lake Hunting Club across the river from Burlington. In the same essay, he explained the constraints his father had placed on him while hunting partridge. "When my father gave me the shotgun, he said I might hunt partridges with it, but that I might not shoot them from trees. I was old enough, he said, to learn wing-shooting. My dog was good at treeing partridge, and to forego a sure shot in the tree in favor of a hopeless one at the fleeing bird was my first exercise in ethical codes. Compared with a treed partridge, the devil and his seven kingdoms was a mild temptation" (121).

Scenes from Leopold's childhood emerge subtly in other essays. In "Goose Music," first published in *Round River,* Leopold describes a boy, strikingly like himself, who was

> brought up an atheist. He changed his mind when he saw that there were a hundred-odd species of warblers, each bedecked like to the rainbow, and each performing yearly sundry thousands of miles of migration about which scientists wrote wisely but did not understand. No "fortuitous concourse of elements" working blindly through any number of millions of years could quite account for why warblers are so beautiful. No mechanistic theory, even bolstered by mutations, has ever quite answered for the colors of the cerulean warbler, or the vespers of the woodthrush, or the swansong, or—goose music. I dare

say this boy's convictions would be harder to shake than those of many inductive theologians.[7]

Leopold's convictions were obvious to his new classmates at Lawrenceville Prep. Within a month of his arrival, his outdoor activities earned him the nickname, "the Naturalist." He came to Lawrenceville shy, earnest, serious, and more than a little judgmental, but his lighter, social side soon emerged and he found his niche in the convivial community of the school. He was not a typical student. He was after all from Iowa, which was infested with grizzly bears and gunfighters for all his eastern mates knew. He quickly became known about the school as a "shark," an ace in the classroom (except in algebra, which eluded him).

Most conspicuous of all was "Leo's" penchant for long hikes ("tramps," he called them) into the New Jersey countryside. Dressed in his knee boots and winter coat, bandanna around his head, plant collection box slung over his shoulder, walking stick in hand, Aldo set forth on tramps to Princeton, Washington's Crossing, Trenton—sometimes as often as four or five times a week, in all manner of weather. He stepped out one day into a driving blizzard. "Progress was in some places impossible," he wrote home, adding that he could "not remember ever enjoying a bit of winter weather more."[8] Over his year-and-a-half career at Lawrenceville he came to know every field and stream and woodland within a ten-mile radius of the school, and to his study of birds he added an equally eager study of the local plant life.

All the while, Leopold recorded his observations in a remarkable series of letters to his family in Burlington. Before Aldo left home, Clara had implored him to write long and often. Aldo responded with enthusiasm. His detailed accounts of his daily tramps foreshadowed the literary expressions for which he would later become celebrated:

I have seen many birds of beauty, but not one of such dazzling plumage and bearing as the Blackburnian. The upper parts and wings are jet black, with various white markings, especially in the wings and tail, the underparts pure white, heavily streaked with black, and the throat, sides of neck, and head, and middle of the black crown, of deep, rich, fiery orange. I have never seen such a color, soft and beautifully offset by the black, yet apparently about to burst into flame, like a red hot coal. The Blackburnian is a prize for rarity alone, but doubly one for his gorgeous plumage.[9]

At the same time, his deep conservation concerns began to surface. When his father wrote and informed him of the annual spring decimation of waterfowl along the Mississippi, Aldo replied, "I am sorry that the ducks are being slaughtered as usual, but of course could expect nothing else. When my turn comes to have something to say and do against it and other related matters, I am sure nothing in my power will be lacking to the good cause."[10]

Aldo took the first step toward fulfilling that prediction when he enrolled in the Sheffield Scientific School at Yale University in 1905, intent on later entering the Yale Forest School, which was then graduating the nation's first trained foresters. Forestry was a booming profession, at the forefront of the broader conservation movement that was flourishing under Teddy Roosevelt's leadership. The Yale School was less than a decade old, having been established with funds provided by the family of Roosevelt's chief forester, Gifford Pinchot.

Aldo spent four years in New Haven, gaining a bachelor of science degree and a master's degree in forestry. Along the way, he would go through a bewildering series of personality shifts. Shedding his youthful shyness, he became known as quite a dandy, sporting fine clothes and courting a number of young ladies from the eastern finishing schools. His tramps persisted into his second year at Yale but quickly dwindled down to a bare few. His interests in the cultural advantages of the east coast and in extracurricular activities came into conflict with his interest in the outdoors and the solitary life it offered. Sleepless nights and overwork ensued. As he began his formal study of forestry, his old perspective as a naturalist fell aside.

Forestry under Gifford Pinchot meant the wise use of forests: the efficient management and development of the nation's public and private forestlands. In the early days of the profession, a disproportionate number of those who would carry out this mission graduated from Yale. The romance of the forester's life was soon lost in the labs and classrooms and Mechanical Properties of Wood, Timber Management, Forest Law, Forest Mensuration,

Forest Insects. Aldo devoted himself to his studies, admitting at one point that he was "getting narrow as a clam with all this technical work." Still, the romance survived. He stubbornly declared that he had "no ambition to be a tie-pickler or a Timber-tester." [11] One day, after a lecture on the organization of the United States Forest Service, Aldo's friend Rube Pritchard exclaimed that he would rather be a supervisor on one of the nation's new National Forests than be king of England. There was not a young forester alive who did not agree.

Carl Leopold taught his son to appreciate the natural world as a sportsman, a woodsman, and a naturalist. Forest School taught him to see it as a professional forester. It gave him the knowledge and techniques to manage forests "for the greatest good for the greatest number over the long run." The days of profligate, indiscriminate destruction of forests were coming to a close as Leopold and his generation of progressive foresters dispersed to the far corners of the continent. The new day would not arrive without backbone, hard bargaining, and a few rounds of gunfire, but for the good of the nation it had to arrive.

There was never a piece of terrain that failed to excite Leopold's interest. Whether viewed on foot, from a train or rowboat or stagecoach, or on horseback, landscapes were always surveyed with a discriminating eye for their history—natural and human—and for their unique flavor. And it never took long for a particular piece of country to exercise its pull on Leopold. He guarded his freedom jealously—that was, after all, one of the motives behind his choice of career—but he easily became attached to new places.

Few places pulled as strongly on Leopold as did the sprawling high country of the Apache National Forest in the Arizona Territory. He arrived there, green as the mountain cienegas, in July 1909. Forest headquarters were at Springerville, on the volcanic southern rim of the Colorado Plateau, between the pinyon-juniper range and the ponderosa pine forest, beneath the brooding hulk of Escudilla Mountain. It was wild country, barely settled, imposing, diverse; a true wilderness of high alpine meadows; wolves, grizzly bears, and deer; great stands of pine and

Aldo Leopold's Early Years

folded recesses of tangled, semiarid canyons. Leopold arrived on the stage out of Holbrook, bought himself a horse and a complete cowboy outfit, and set about lightening his deep eastern shade of green.

After just a month on the Apache as forest assistant, Leopold was assigned to lead a timber reconnaissance team across the rugged upper canyons of the Blue River. It was apparently here that he and his fellow cruiser shot the mother wolf that figures so prominently in "Thinking Like A Mountain."[12] But it would be years, and many an antiwolf campaign later, before he would come to regret the incident. Pioneers did not pause to bemoan the loss of a "varmint."

Leopold botched that first reconnaissance assignment badly, enough to prompt an investigation by the regional office of the Forest Service in Albuquerque. He had mismeasured an important baseline, mismanaged the camp, imposed his own demanding standards of outdoor etiquette on his crew, and ignored the experience of veteran timber cruisers in the team. It was the first of many ego-deflating experiences that he would have to endure before coming to terms with the landscape and the job before him.

Leopold put in two years on the Apache—exciting, colorful years that never lost their freshness in his mind. Three decades later he would write about them ("On Top," "Escudilla," "Thinking Like A Mountain") with a combination of celebration, sorrow, and regret: celebration of a land of wonder, sorrow that some of that wonder had disappeared, and more than a little sad recognition that "we forest officers" had had a hand in that disappearance.

Leopold's rise in the Forest Service was rapid. In the spring of 1911 he was assigned to the Carson National Forest in northern New Mexico as deputy supervisor. The Carson was among the district's most beautiful forests, climbing the heights of the Sangre de Cristos in the east and spreading out over the San Juan Mountains to the west. It was also one of the most abused forests in the country. Decades, even centuries, of heavy grazing pressure on the upper Rio Grande valley had bequeathed a ruined range, and it was up to the Forest Service to take control.

Aldo quickly became versed in the details of range management and range politics. He and his fellow officers quite literally stuck to their guns, and reform came, grudgingly but effectively, to the Carson. In 1913, Leopold was named supervisor of the Carson, the highlight of his young career.

In the meantime, he had more on his mind than grazing permits and sheep driveways. He had met and fallen seriously in love with Estella Bergere, a daughter in one of New Mexico's most prominent and powerful families. Her mother was one of the celebrated Lunas, a family whose remarkable lineage traced back to the nobility of medieval Spain and whose reigning patriarch, Don Solomon Luna, controlled one of the nation's great sheep empires. Estella was a twenty-one-year-old school teacher when she and Aldo were first introduced at a drug store in Albuquerque. She was as gracious and lively as any young lady in New Mexico—including her six equally charming sisters. Aldo was immediately taken with her character and her distinctive Spanish features. Estella was slender and of average height, with dark hair (that had, according to Aldo, "a reddish glint should you ever see it exactly right"),[13] deep brown eyes, a long aquiline nose, and olive skin. By their third meeting, Aldo was in love.

Courtship was no easy matter when one's job took one to distant mountains and when one's beloved lived seventy miles, a mountain range, and an unreliable railroad line away. Aldo, however, was persistent, making effective use of the mail run, disarming Estella with his passionate letters and ardent professions of love. Taken aback at first, she found her own love growing steadily. Aldo proposed to her in August 1911, just four months after they met. In November she said yes, and they were married on October 9, 1912, at the Cathedral of Saint Francis in Santa Fe.

The next thirty-six years of their lives were marked by a mutual, utter devotion to one another. Theirs was an old-fashioned marriage, tender and plainly fulfilling to both of them. Estella was a pious Catholic and, faithful to Spanish tradition, saw to it that her husband and children were provided with a good home. Without her, Aldo simply could not have accomplished what he did. Far more than the keeper of his home, she was his closest critic and listener, never wavering in her devotion to him nor in her faith in his abilities. She was the center of Aldo Leopold's life.

Six months after he and Estella married, Aldo was caught in a spring snowstorm that nearly cost him his life. The blizzard came on while he was riding on inspection in the San Juans, and he was forced to spend several miserable nights in the high country along the continental divide. By the time he finally reached forest headquarters at Tres Piedras, he was fearfully swollen, a victim of acute nephritis. He made it to a doctor in Santa Fe with only hours to spare.

The attack kept him away from work for a frustratingly long sixteen months. Restricted in diet and ordered not to exert himself, he languished in Santa Fe and Burlington, unable to do anything but read, write occasional letters, and visit with friends and family. When he finally did return to the Forest Service in October 1914, it was not to the Carson National Forest, but to the District Office of Grazing in Albuquerque. He pushed paper: grazing claims, reports, applications, complaints, appeals.

In the meantime, however, he had renewed his devotion to the cause of wildlife conservation (or, more properly, to the antecedent campaign for "game protection"). Prompted by his enforced period of reflection, his reading, and perhaps by the death of his father in December 1914, Leopold began to address the issue from his position within the Office of Grazing. His boss, District Forester Arthur Ringland, knew Leopold's enthusiasm for the game conservation work and created a new position in which he could devote to it his full energies.

Leopold's new job also gave him responsibility for the district's publicity and recreation work, but game protection was his foremost concern. In the fall of 1915 he met William Temple Hornaday, the nation's best-known and most zealous game protector. In the months that followed, Leopold took his campaign around the state and became the driving force behind the New Mexico Game Protective Association (NMGPA), a statewide organization of sportsmen devoted to the reform of the state's fish and game administration. Over the next four years, he worked tirelessly to promote the NMGPA's program, which called for new game laws, better law enforcement, the establishment of a nonpolitical state agency, the creation of game refuges, and an intensified effort to "control" predators. To push for these measures and to unify his group, Leopold helped write and edit a newspaper, the *Pine Cone*, which attracted regional and even national attention. As the tide turned in New Mexico, Leopold gained recognition in national conservation circles and received in 1917 the Gold Medal Award of Hornaday's Permanent Wild Life Protection Fund.

Aldo and Estella had a family by now. Between 1913 and 1919, Estella gave birth to four children: Starker, Luna, Adelina, and Carl. Aldo began to write more during these years, publishing the first of the more than three hundred articles of his career. Though still weakened by his illness, he began to hunt again, slowly at first, but increasingly as his sons grew. He began the meticulous journal-keeping that he continued for the rest of his life. Aldo was also becoming one of the best-read foresters in the country. In the evenings, he and Estella often read to one another, a lifelong habit that gave Aldo wide familiarity with history, natural history, literature, philosophy, and the classics. All came to bear in his thinking and writing on conservation.

Leopold left the Forest Service in 1918 for a brief stint as secretary of the Albuquerque Chamber of Commerce, thinking that this would give him greater opportunity to promote game protection. Already, however, he had begun to think in terms of actual management, not just protection, of game in order to increase populations. Such a science had been theorized, but in reality it had not advanced beyond captive breeding and release programs—a far cry from the goal of increased production of game in the wild. Much of Leopold's writing at the time discussed the possibilities of such game management within the context of forestry.

Leopold returned to the Forest Service in 1919 as chief of operations, the second highest position in the district. This put him in charge of a broad range of forest activities, from personnel and fire fighting to telephone-line installation and ranger-station construction. Not everyone in the district appreciated the choice of Leopold for the job—not even his new boss. Although thoroughly competent, he was suspect on account of his high profile and his unorthodox ideas. Many felt as well that he had been advanced beyond his experience.

Leopold took the protests as a challenge. Over the next five years, from 1919 to 1924, he blazed a wide trail of innovations in forestry policy and administration. He reformed the system of inspection that was at the core of operations work. He became the chief advocate and leading spokesman for a group that wished to set aside parts of the forest lands as wilderness areas. The so-called good roads movement had overtaken the Forest Service and was reducing roadless areas in the United States dramatically. In an attempt to retain a few large areas for recreational purposes, Leopold wrote several seminal articles that argued for "some logical reconciliation between getting back to nature and preserving a little nature to get back to."[14] This effort culminated in the establishment, in June 1924, of the Gila Wilderness Area on the Gila National Forest, the first such area to be so designated. Leopold's own passion for wilderness travel took him into the Gila several times and also, in 1922, into the backwaters of the Delta of the Colorado River with his brother Carl, a trip Leopold later recalled in the essay "The Green Lagoons."

During these years, Leopold also undertook a less-heralded but equally significant personal effort to understand the delicate equilibrium of the southwestern range. The effort exemplified Leopold's restless curiosity, his ability to correlate facts, his talent for "reading" landscapes. Drawn at first by the intractable problem of soil erosion, Leopold devoted one inspection trip after another, in national forests across New Mexico and Arizona, to understanding the cause and effect that related erosion to overgrazing, flooding, vegetation changes, fire, forest reproduction, siltation rates, and climate. The result was a series of articles and speeches noteworthy for ecological insights decades ahead of their time.[15]

Altogether, Leopold spent fifteen years in the Southwest. When he left in 1924, he was not the same forester that had arrived in 1909. He had never been a full convert to utilitarian thinking in forest management. He could tally board feet with the best of them, but as early as 1913 he was professing the idea that the ultimate measure of success had to be "the effect on the forest." "We are entrusted," he wrote to his men on the Carson at that time, "with the protection and development, through wise use and constructive study, of the timber, water, forage, farm, recreative, game, fish, and aesthetic resources of the areas under our jurisdiction." [16] Long before "multiple use" became the byword in forestry, Leopold endeavored to take a total view of the forest, and to gauge policy accordingly. By the time he left the Southwest, he was increasingly disturbed by the manner in which raw utilitarian motives—"economic determinism" he called it—were coming to dominate development in an environment that was intrinsically sensitive to exploitation and susceptible to damage.

In June 1924, a transfer brought Leopold to the U.S. Forest Products Laboratory in Madison, Wisconsin. Leopold's superiors in Washington recognized his aptitude for research and appointed him assistant director of the lab, anticipating that he would soon rise to director. The current director was talking of leaving the job, and Leopold accepted the position with this in mind.

As it turned out, the director stayed and Leopold remained locked in as assistant for four years. The work of the lab was exclusively technical, but Leopold, obviously "a fish out of water," as one colleague put it, made the most of the situation. [17] He picked up a sound knowledge of wood technology. Conversely, he tried to stretch the purview of the lab, bringing it into closer cooperation with the national forests, encouraging research into wood wastage and the use of inferior timber species.

His most creative work in these years, however, was avocational. In a spurt of articles in 1924–26, he continued to raise the argument for wilderness preservation and became known nationally, particularly within the Forest Service, as the leading advocate in the cause. He also wrote at length on game and game management, laying the foundation for his subsequent change of occupation. It was during the Forest Products Lab years that Leopold and his growing family (a fifth child, Estella, arrived in 1927) took up an interest in archery. In the dank basement of their house on Van Hise Avenue, Aldo carefully crafted the bows, arrows, nocks, quivers, even the bowstrings, he used in his new hobby; Estella usually sat alongside the workbench and knitted. On the archery range, however, Estella proved to be the master. She became Wisconsin's women's champion for five years running.

By 1928, Leopold was well ready to leave the lab. The new directions he tried to encourage at the lab were neglected. His paperwork only piled higher with time; the director showed no inclination to leave. In mid-1928 Leopold left the Forest Service to take a position with the Sporting Arms and Ammunitions Manufacturers' Institute. SAAMI hired Leopold to conduct a national survey of game conditions and to oversee a series of funded game management research projects. The job was an immense and unprecedented one. Hard research on game and game habitat was almost entirely lacking in the United States. Information was largely a matter of myth, local tradition, and unproven hunches. Actual management of game had been developed formally in just one locale, at Herbert Stoddard's quail demonstration area in Georgia. What "management" existed still consisted of captive breeding and release, not of the habitat development that Leopold envisioned.[18]

From 1928 to 1931, Leopold traveled across nine states gathering hard information on everything from game laws to farming practices, geology to hunter behavior. He delivered an important series of lectures in 1929 at the University of Wisconsin on the theory behind game management. He became involved in a national effort by sportsmen to draw up a game policy; the resulting document, written largely by Leopold, would guide work in the wildlife profession for forty years. In 1931, Leopold's *Report on a Game Survey of the North Central States* was published, a landmark study that, for the first time, gave the theory of management a regionwide basis of facts on which to stand.[19]

In his travels across the Midwest, several points became clear to Leopold. First, although game populations were suffering from overhunting, a far more important factor was the destruction of habitat. The ideal of "clean farming" wreaked havoc not only with the farm economy (and, as would soon become clear, with the soil itself), but destroyed the coverts that small and upland game needed. Second, Leopold was now convinced that predators played only a minor role in game depletion. His views on predators had changed drastically since his varmint-control days in the Southwest. In the mid-twenties, he had begun to admit their value to science. By 1930, he was beginning to appreciate their ecological value, having been shaken by the tragic fiasco of deer overpopulation following predator extirpation on the Kaibab Plateau in Arizona. Finally, and most significantly, the game survey convinced Leopold that the most effective agent of game conservation was, and had to be, the landowner, the farmer. This became one of the pillars of Leopold's philosophy. He would always remain skeptical of large-scale government efforts to solve widely dispersed conservation problems, not as a consequence of any strong ideological opinion, but as a matter of practicality.

In 1932, the depression reached Leopold. His support from SAAMI was cut back and then discontinued. With a wife and five children to support, Leopold was in a precarious position. He was a nationally recognized expert in a field that hardly existed. He managed to hold on with a number of consulting jobs in Iowa and Wisconsin, but these were temporary. In his slack periods, he put the finishing touches on his magnum opus, *Game Management*. Published in 1933, it was the first text in the field and provided the fledgling science with its foundation of theory, technique, and philosophy. It would remain the wildlife profession's standard text for several decades.

Soon after the New Deal architects took office, Leopold returned to the Southwest as a consultant to the Forest Service, supervising the work of the Civilian Conservation Corps on a number of erosion control projects. During this period, in May 1933, Leopold delivered an address he called "The Conservation Ethic," an early attempt to define the ideas that ultimately came together in "The Land Ethic" in *A Sand County Almanac*. Fourteen years would pass before Leopold composed the final essay, adding a depth of scientific insight and sense of urgency that were not yet present in 1933, but his fundamental thesis would remain the same: the time had come to regard land as more than mere property, and to exercise ethical restraint in our relations with it.

Aldo Leopold's Early Years

While in the Southwest, Leopold received word that the University of Wisconsin had approved a plan to appoint him professor of game management within the College of Agriculture's Department of Agricultural Economics. Leopold was close to many of the university's officials and faculty, and such a plan had been discussed for a number of years. Still, game management was hardly a respectable science at the time. The antipathy came from both sides: scientists looked down upon it as a mere adjunct to blood sport, while many hunters, and even game officials, saw it as mere impractical theorizing. Leopold would make the program work through the breadth of his vision. He already saw that game management constituted a vital link between the economically oriented conservation in which he had been trained and the ecologically based conservation toward which the movement was heading. His program was funded on a five-year experimental basis and in the fall of 1933 he joined the University of Wisconsin.

Leopold would remain at the university for the remaining fifteen years of his life. These would be his most productive years. "The Professor," as his students called him, would gain a reputation as a penetrating teacher, uncommonly kind but highly demanding. Several generations of students, graduates and undergraduates alike, would absorb his techniques, his philosophy, and most of all his way of seeing land as history, as habitat, as a living, ever-changing entity. He would continue to write prodigiously, to guide the science he had helped invent, to serve Wisconsin as a teacher and a Conservation Commissioner, and to work in dozens of local and national conservation organizations, from the Wilderness Society down to the Riley Game Cooperative in western Dane County. And, beginning in 1941, he would begin work on the collection of essays that ultimately became *A Sand County Almanac*.

There is no "Sand County" in Wisconsin. Leopold had visited the central sands region of Wisconsin numerous times prior to 1935, on hunting and fishing trips, during his game survey, in his consulting work with the state. In 1933 and 1934 he returned on a number of occasions in his cooperative work with the New Deal

CURT MEINE

agencies that were setting up their tents there. Leopold was no New Dealer himself, but he was often called in for consultation on their plans.

1934 was a difficult year in the central sands, for people and land. Drought and dust storms visited in the spring, drying up and carrying the thin topsoils off of several of the vulnerable sand counties. In the dessicated low meadows and the man-drained marshlands, fires smoldered in the peat. And over all of this, the veil of the depression still cast its dulling pall. The spring drought did break, and the dust did settle, but the local conditions would prove to be sadly prophetic of the dust storms then gathering on the high plains of the West.[20] The remainder of the 1930s would add a grim, grit-laden resolve to the conservation call.

One day during that summer of 1934, Aldo and his brother Carl were returning from a fishing trip in Waushara County, one of the sand counties. On the way back to Madison, they stopped to investigate a report that Leopold had received of a nesting pair of sandhill cranes in the Endeavor Marsh near Portage. Leopold estimated at the time that there were no more than twenty breeding pairs in the entire state. The exploitation of its marshland habitat had endangered its local existence. Aldo and Carl talked to a farmer whose fields backed up against the marsh. The farmer not only told them where they might find the pair, but gave them a lengthy history of the farm, the cranes, and the marsh. Hiking out along some high ground, they neared an oak, and then spied the tall, gray birds on the marsh edge—but only for a moment. The wary birds lifted themselves up over the cat-tails and disappeared into the more isolated marsh interior.[21]

The sighting captivated Leopold and symbolized in many ways the significant changes that were occurring in his thinking about man and land. He returned to Madison and began to delve into the lore and biology of the crane. He consulted his friends in ornithology and returned to the marsh for longer ob-servation. Three years later, he would write the poetic celebra-tion of the crane and its haunts, "Marshland Elegy."

By the time Leopold wrote "Marshland Elegy," *game manage-ment* had broadened into *wildlife management*, and Leopold himself was thinking in yet broader terms, as a land ecologist and manager. In this, of course, he was not alone. The late 1930s saw a tremendous cross-pollination of the sciences underlying conservation, and Leopold was but one of many voices to emerge from the times. The authority, humor, and clarity of that voice, however, were rare. And much of its strength is due to the fact that, in the winter of 1935, Leopold found a home for his curiosities on the south end of Wisconsin's sand counties.

It was, and is, just a piece of land—less productive, less scenic, less wild, less "natural" even, than a thousand other places in Wisconsin alone. But in country, as in people, a plain exterior often conceals hidden riches, "to perceive which requires much living in and with."

Of what account is this? To what end perception? Why need we look to land when, in Leopold's words, "education and culture have become almost synonymous with landlessness"? [22]

Leopold's perception, like other human accomplishments, is admirable in and of itself, but it also offers us a sense of our own potential. *A Sand County Almanac* allows us to see the world as Aldo Leopold saw it. His view was unique, extraordinary, in some ways extreme. Perhaps great accomplishment in any field of endeavor requires eccentricity in this best and literal sense of the word. Through his eyes, we learn to think about and appreciate land for ourselves.

A generation has passed since *A Sand County Almanac* first appeared. Today's conservationists carry on the work of the previous generation, and another generation prepares to assume the mantle. And so, for the good of both man and land, the work must continue. The conservation imperative becomes only more important as we come to understand that freedom is a function of perception; that we are free to the degree that we are aware of, understand, and can respond to the forces that shape our lives. Ignorance of environmental processes, in partnership with a human economy that sees land only as a giver and not a receiver of gifts, can in the long run only impoverish and constrain our lives, and the life around us. It is perhaps the greatest contribution of Leopold and his compatriots that they lead us to the truth that freedom itself is a kind of endangered species, one that suffers or prospers depending on the relationship between the environment and those who inhabit it. Leopold allows us to see and understand the dynamic force "which underlies and conditions the daily affairs of birds and men" (96). In so doing, in telling us his monthly stories of the life on a reborn farm in central Wisconsin, Leopold served and preserved freedom in its most comprehensive, uplifting sense.

Notes

1. Aldo Leopold, *Round River* (New York: Oxford University Press, 1953), 151.
2. Background material for this discussion of Leopold's family history is drawn primarily from interviews with Aldo Leopold's youngest brother, Frederic Leopold of Burlington, Iowa. Also helpful were Frederic Leopold's unpublished manuscripts, "Leopold Family Anecdotes" (1982), "Aldo's Middle School Years—Summer Vacation" (undated), and "Recollections of an Old Member" (undated).
3. "Business Men's Resolutions of Respect," 13 February 1900, in unclassified folio, Leopold Papers (LP), University of Wisconsin–Madison Archives.
4. Frederic Leopold, "Leopold Family Anecdotes," 21.
5. In Leopold's school composition book, essay on "Birds," LP 12B3.
6. This trip to Yellowstone took place in September and October of 1903. Related materials may be found in LP 8B2,4.
7. Aldo Leopold, *Round River,* 171.
8. Aldo Leopold to Marie Leopold, 25 January 1905, LP 8B4. This account of Leopold's school years is based on the many letters he wrote during the years 1904–1909, LP 8B4–6.
9. Aldo Leopold to Clara Leopold, 6 May 1904, LP 8B4.
10. Aldo Leopold to Clara Leopold, 21 March 1904, LP 8B4.
11. Aldo Leopold to Clara Leopold, 21 February 1909, LP 8B6.
12. Evidence of this date is circumstantial. Documentary evidence from Leopold's years in the Southwest is uneven, consisting of letters, Forest Service diaries, hunting journals, publications, and assorted unpublished notes, manuscripts, and reports. An event as significant as the killing of a wolf could hardly have escaped Leopold's record keeping. Nevertheless, the only extant evidence of the incident is in the essay "Thinking Like a Mountain" itself. A process of elimination suggests

that it took place in the fall of 1909, shortly after Leopold's arrival on the Apache National Forest. Leopold's Forest Service diaries give a complete record of his two years on the Apache, except for this period. His diaries from this several-month period were requisitioned during the investigation of Leopold's conduct on the 1909 reconnaissance.

13. Aldo Leopold to Clara Leopold, 7 June 1911, LP 8B7.

14. Aldo Leopold, "The Wilderness and Its Place in Forest Recreational Policy," *Journal of Forestry* 19, no. 7 (November 1921): 718.

15. See especially Aldo Leopold, "A Plea for Artificial Works in Forest Erosion Control Policy," *Journal of Forestry* 19, no. 3 (March 1921): 267–273; "Pioneers and Gullies," *Sunset Magazine* 52, no. 5 (May 1924): 15–16, 91–95; "Grass Brush, Timber, and Fire in Southern Arizona," *Journal of Forestry* 22, no. 6 (October 1924): 1–10.

16. Aldo Leopold to officers of the Carson National Forest, 15 July 1913, in the Carson *Pine Cone,* July 1913, LP 11B1.

17. L. J. Markwardt, interview with author.

18. For a general history of wildlife management in the United States, see J. B. Trefethen, *An American Crusade for Wildlife* (New York: Winchester Press and the Boone and Crockett Club, 1975).

19. Aldo Leopold, *Report on a Game Survey of the North Central States* (Madison: Democrat Printing Company, for the Sporting Arms and Ammunition Manufacturers' Institute, 1931).

20. For an account of the 1934 drought in central Wisconsin, see Michael Goc, "The Great Dustbowl: It First Came to Wisconsin," *Wisconsin Trails* 25, no. 4 (July–August 1984): 20–24.

21. Aldo Leopold, "Wisconsin Journal, 1934–1935," 20 July 1934, LP 7B2.

22. Aldo Leopold, *Round River,* 155.

2 ✶ Aldo Leopold's Sand Country

SUSAN FLADER

We are interested in the convergence of a person and a place—a person who has come to be regarded as a prophet in the evolution of a new relationship between man and land, and a place little known and undistinguished save as it stimulated in that person and others a heightened sensitivity and deepened respect for the larger community of life. It was a coming together in maturity. Aldo Leopold was well past the midpoint of his life when he came to the sand area of central Wisconsin. And the sand country was older still, its long history intimately bound with the geologic and human history of the continent. Leopold encountered the sand counties first as a professional land manager, seeking through new tools of public policy and new techniques of resource management to redress the balance of nearly a century of frontier exploitation. His acquisition of a worn out, abandoned sand farm in 1935 initiated a different relationship with the land, at once more personal and more universal. From his own direct participation in the life of the land he came to a deeper appreciation of the ecological, ethical, and esthetic dimensions of the land relationship.

The sand country of central Wisconsin is a flat land, and old. It is a legacy of sands slowly settling in the shallow Cambrian sea

This essay, abridged, revised, and annotated for this volume, was originally entitled "The Person and the Place" and published in Charles Steinhacker with Susan Flader, *The Sand Country of Aldo Leopold* (San Francisco: Sierra Club, 1973): 8–49.

that covered the interior of the continent half a billion years ago. The sands were compacted, cemented, and capped by limestones and dolomites during subsequent transgressions of the Paleozoic seas, only to be exposed by intervening eons of erosion, then reworked and redeposited by meltwaters of the great continental glaciers.[1]

We know most about the last glacial incursion—the Wisconsin stage—which began seventy thousand years ago and lasted sixty thousand, a series of advances and retreats of a multilobed ice sheet. It was the Green Bay lobe of the Cary substage of the Wisconsin glaciation, fed by the snows of Labrador, that moved down out of the northeast a mere fifteen thousand years ago and molded much of the character of this region. The ice stopped in its advance about a third of the way across the sand counties, dumping its load of gravel and rock in a north-south heap.

When the debris-laden ice dammed up the outlet of the Wisconsin River through the Baraboo Hills in south central Wisconsin, it created a huge lake, Glacial Lake Wisconsin, over eighteen hundred square miles covering parts of five counties. Glacial meltwaters rushing into this lake from the north and east dropped their gravels near shore and sent their loads of sand and clay in suspension far out across the bed of the lake. Waves pounded away at the outlying islands of quartzite and sandstone, adding still more sands to the lake and leaving isolated buttes, mounds, castellated bluffs, and pinnacles rising from a level marshy plain when the waters receded.

Glacial Lake Wisconsin was not long-lived. The halting of the ice was a sign that the climate had already moderated and the ice would soon begin its retreat. Meltwaters that filled the lake began draining to the northwest and then later moved in the old southerly direction. The new Wisconsin River rapidly cut itself a gorge through the weak Cambrian sandstone at the now-heralded Wisconsin Dells and found a new course around the east flank of the Baraboo Hills across the flat swampy plain known from the time of the earliest French explorers as The Portage. Here the Wisconsin River on its way to the Mississippi flows within a mile and a half of the Fox River, which drains northeastward to Green Bay and the St. Lawrence.

Central Wisconsin is richly diverse—despite or perhaps because of the poverty of its sands—owing in large part to its position at a boundary between two floristic provinces: the prairies, oak savannas, and southern hardwoods to the south—and the pine savannas, conifer-hardwoods, and elements of boreal forest in the north. The preeminent student of the vegetation of Wisconsin, Leopold's friend and colleague, John T. Curtis, described this boundary as a "tension zone"—a band ten to thirty miles wide running diagonally southeast across the state and marking both the northern limit of many southern species and the southern limit of many northern species. Some botanists consider the entire sand area as a widened part of the tension zone, its sands accentuating differences in moisture content and temperature, its low-lying glacial lake beds and river bottoms more prone to frosts, its nutrient-poor soils preventing any one species from taking over. The result is an intermingling of species from north and south.[2]

The woodland peoples who inhabited the area from before the time of Christ until the coming of the French explorers were mostly hunter-gatherers who lived in small nomadic bands that followed the animals from day to day, season to season, century to century. They also fished, made maple sugar, and gathered wild rice; and some of them—especially during periods of more favorable climate—raised corn, beans, and squash in small garden plots along the rivers. It is possible that the central sand area with its prairie-forest transition belt may have been so rich in animal life and so strategically located that it could not be held as the exclusive terrain of any one group of native people, but rather served as the hunting grounds of several different groups and as a neutral buffer zone between them.[3]

For all their importance to the Indians, the hunting grounds of central Wisconsin remained unknown to the whites for nearly two centuries after Jean Nicolet greeted the Winnebagos in 1634 on the shores of Green Bay. Louis Jolliet and Father Jacques Marquette made the first officially recorded traverse of Wisconsin via the Fox-Wisconsin waterway in 1673, and thus became the first whites to describe the area. Yet they were only skirting the edges of the sand country.

Aldo Leopold's Sand Country

Within a decade both Robert Cavelier, Sieur de La Salle, and Daniel Greysolon, Sieur du Lhut, had trading parties on the Mississippi and were contending for control of the Fox-Wisconsin route. La Salle had arrived by way of Chicago and the Illinois River and du Lhut by way of the Brule–St. Croix route from Lake Superior. But the easiest route of all between the Great Lakes and the Mississippi—passable by canoes at high water even without unloading—was the Fox-Wisconsin link forged by the Green Bay lobe of the Cary ice. As the glacier butting against the Baraboo Range had left the sand country in its backwaters, so also did the Fox-Wisconsin portage route, famed passageway of explorers and fur traders for nearly two centuries.

The change came with white settlement. Until 1820 the lands west of Lake Michigan were Indian country. In the 1820s, however, whites pushed into southwestern Wisconsin, lured there by rich veins of lead. The Winnebago, Chippewa, Ottawa, and Potawatomi ceded lands in the lead region in 1829, and the Black Hawk massacre of 1832 persuaded Indians to turn over the rest of their holdings south and east of the Wisconsin and Fox rivers. Government surveyors worked the area during 1832–36 and within a decade most of the best lands were taken up by settlers or speculators. Winnebago, Menominee, and Chippewa Indians meanwhile retreated to more remote portions of their former territory in central and northern Wisconsin, until white entrepreneurs, stimulated by the heavy demand and high price of lumber in the newly opened prairie counties to the south, initiated a logging boom on the Wisconsin River. The Indians were forced to cede their remaining lands, save for a few small reservations. But the Winnebago, rounded up four times by soldiers, kept returning to their sand country hunting grounds.

Pioneer settlers moved into central Wisconsin after the last Indian cession in 1848, the year Wisconsin achieved statehood. The first comers claimed the better soils of the glaciated prairies and oak openings to the east; later arrivals had to make do with the successively sandier, less fertile lands farther west.

Among the earlier arrivals were Daniel Muir and two of his young sons, John and David, who had left Dunbar, Scotland, for the New World in February 1849. Daniel Muir selected a quarter section of land—one hundred and sixty acres—four miles from the nearest neighbor in the sand country north of The Portage, and quickly built a shanty of oak logs.

"To this charming hut, in the sunny woods, overlooking a flowery glacier meadow and a lake rimmed with white water-lilies," John Muir reminisced more than half a century later, "we were hauled by an ox-team across trackless carex swamps and low rolling hills sparsely dotted with round-headed oaks." No sooner did they arrive than the boys discovered a blue jay's nest, complete with green eggs and beautiful birds, and then a bluebird's and a woodpecker's nest, and frogs and snakes and turtles— "Oh, that glorious Wisconsin wilderness!" They marveled at the newness and freshness of it all, at the "extravagant abounding, quivering, dancing fire" of millions of lightning bugs on a sultry evening, the 'boomp, boomp, boomp" of partridge drumming, the mysterious winnowing of jacksnipe spiraling high overhead and plummeting earthward, the great long-legged sandhill cranes on the meadow. "No other wild country I have ever known extended a kinder welcome to poor immigrants," Muir remarked.[4]

But there was work to be done, the hard work of making a farm, and Daniel Muir was a dogged pioneer. There were trees to chop and brush to cut down and burn, tough prairie sod to turn with the plow—taking care to dodge all the stumps that later would have to be dug and chopped out to make way for the McCormick reaper. There were wheat, corn, and potatoes to plant and then to hoe, stovewood to chop, fence rails to split, and a new frame house to finish before the rest of the family arrived from Scotland in the fall. Then the harvest: cutting, shocking, and husking the Indian corn, cradling, raking, and binding the wheat, stacking and threshing. "It often seemed to me," wrote John Muir, "that our fierce, over-industrious way of getting the grain from the ground was too closely connected with grave-digging. . . . Men and boys, and in those days even women and girls, were cut down while cutting the wheat." He was convinced that all the plowing, chopping, splitting, grubbing, hoeing, and threshing of his childhood stunted his growth. Why toil and sweat and grub oneself into an early grave on a quarter section of land, trying in vain to get rich, he wondered, when a comfortable enough living could be won on a fourth as much land, and time could be gained to "get better acquainted with God" and maybe even enjoy life?[5]

To get cash for wheat—fifty cents a bushel—it had to be hauled by wagon to Milwaukee, a hundred miles away. In the first years they got up to twenty-five bushels an acre, but within five years the soil was so exhausted that they got only five or six bushels even in the better fields. Yet Daniel Muir doggedly cleared more fields, planted them to wheat and then to corn, built more fences and outbuildings for cattle and pigs and horses. And then, after eight years of effort, "after all this had been victoriously accomplished, and we had made out to escape with life," John Muir recalled, he bought a half-section of wild land five miles away and started all over again.[6] To what end?

It was a poor land for making money, as John Muir remembered it, but rich country for living.

Shortly after John Muir left the sand country in 1860 to exhibit his hand-carved hickory clocks at the state agricultural fair in Madison and to make his own way in the world, the most noted historian of the pioneer experience was born in the bustling river town of Portage. The Wisconsin of Frederick Jackson Turner's boyhood was a rapidly evolving social order. As Turner roamed the streams and lakes of the sand country in quest of trout or pickerel or bass, he occasionally chanced upon Indians who had escaped their seemingly inevitable fate of removal to reservations. He also observed later stages of succession as the forests gave way to wheat and as the earliest white settlers moved on west, nudged along by newer arrivals from Germany, Scandinavia, and Britain. In town the changes were even more dramatic—the coming of the railroads, the erection of new commercial establishments and public buildings, the rapid population turnover, and the medley of ethnic groups that made local politics ever more complex and fascinating. Everywhere was expansion.

From a sense of being part of a society oriented to the future, Frederick Jackson Turner was to forge a stunning explanation of the American past. The Turner frontier thesis, as it is called— after Turner's most important statement of the theme in an 1893 address, "The Significance of the Frontier in American History"—dominated American historical thought for the first quarter of the twentieth century, spawning a "frontier school" of scholarship. Since then the thesis has been subjected to repeated criticism and revision, but the frequency and intensity of the discussion is in itself tribute to the resonant chord Turner struck in the American mind.

"The peculiarity of American institutions," Turner pointed out, "is the fact that they have been compelled to adapt themselves to the changes of an expanding people—to the changes involved in crossing a continent, in winning a wilderness, and in developing at each area of this progress out of the primitive economic and political conditions of the frontier into the complexity of city life." The driving force of American development was the existence of an area of free land, and the existence of free land exerted a transforming influence not only on the frontier but on the entire nation and even on the Old World.[7]

Turner was telling Americans what they already intuitively knew about themselves. Thomas Jefferson had stated the belief of an earlier generation that a nation of independent farmers was the greatest assurance of democracy and that a happy republic, free from the social strife of Europe, would endure so long as there was the possibility of expanding to new lands. At the same time he realized that the very existence of free lands, especially in view of the scarcity of capital and labor, made destructive exploitation of the land almost inevitable. There was a paradox here; but Americans, if they thought about it at all, were characteristically willing to charge off the costs of exploitation against the benefits of individualism and democracy and national pride.

But now the frontier was drawing to a close. According to a report of the U.S. Census of 1890, it was no longer possible to draw a line of frontier settlement on the map. Whether one affirmed the pioneer values, as Turner did, or questioned them, perhaps in the manner of John Muir, there was still the nagging problem: What happens to America when there is no more free land?

Turner called it the problem of the West: "A people composed of heterogeneous materials, with diverse and conflicting ideals and social interests, having passed from the task of filling up the vacant spaces of the continent, is now thrown back upon itself, and is seeking an equilibrium. . . . The forces of reorganization are turbulent and the nation seems like a witches' kettle." What was the nature of the equilibrium? The furthest he would venture, in a rather remarkable bit of provincialism, was to suggest that the Old Northwest—Turner's Wisconsin and neighboring states—held the balance of power and was the battlefield on which the issues of American development would be settled. But the problem of the West was not easy of solution—it meant "nothing less than the problem of working out original social ideals and social adjustments for the American nation."[8]

Thus, ironically, through Turner's frontier thesis, the sand country of his boyhood—backwater of continental glaciation and exploration, zone of contention in the agelong battle between prairie and forest, buffer between tribe and tribe—entered center stage in the nation's history.

After his rise to fame, Frederick Jackson Turner spent part of his career as a professor of history at Harvard. But in June 1924, when he retired from teaching, he returned to the University of Wisconsin to continue his research and writing. Also moving to Madison that summer of 1924, settling by chance in the gray stucco house just two doors west of Turner's Cape Cod bungalow, was another interpreter of the American past with a concern for the shape of its future. Aldo Leopold would experience the sand country at the end of his life, as Turner and Muir did at the beginning of theirs, and draw from it values for a new generation.

Although Leopold made a number of hunting and fishing forays into the sand counties during his first years in Madison and later reported on the region in his *Report on a Game Survey of the North Central States* (1931), it was not until his appointment in 1933 to a new chair of game management at the University of Wisconsin that he began to focus sustained attention on the area. Established in the depths of the depression, the chair was lodged in the university's Department of Agricultural Economics in anticipation of Leopold's work on problems of land utilization on Wisconsin's cutover, tax-reverted, burned out, and eroded lands. The sand counties were a case in point.

Settled in the 1850s, '60s, and '70s during the rapid plunder of the central Wisconsin pines, the sand counties had done surprisingly well by their inhabitants as long as the lumber camps lasted. The camps provided ready markets for farm produce and winter jobs for the industrious. But the timber boom passed with the pine, and the soils would not sustain the wheat that Daniel Muir and Frederick Turner viewed as the area's future. On the glaciated sands of the eastern counties farmers shifted increasingly to dairying, at best a marginal operation in competition with more favorable regions of the state; and at worst, where the sand was loose and too many cows overgrazed the scant vegetation, a "blow-out" to both men and land. West of the moraine, blowouts came even faster and the shifting dunes grew with the years. Large portions of the central Wisconsin cutovers were never plowed at all, but devastated by repeated slash fires that killed new seedlings of white and red pine and encouraged the spread of jack pine and scrub oak "barrens."

Out in the marshes, where land could be bought for fifty cents an acre, farmers from the neighboring hills cleared ever more tamarack, or let fire do it, and each summer cut and stacked the new lush growth of wild hay. To Leopold looking back, these haymeadow days seemed the Arcadian age for marsh dwellers: "Man and beast, plant and soil lived on and with each other in mutual toleration, to the mutual benefit of all. The marsh might have kept on producing hay and prairie chickens, deer and muskrat, crane-music and cranberries forever" (99). Such had been John Muir's vision too, but the ambitions of industrious pioneers like his father dictated a different future for the haymarshes. In the dry years of the early 1890s farmers anxious to get ahead tried plowing the haylands for crops and were rewarded with the bountiful yields of any virgin soil. When rains returned to thwart their ambitions, the land boomers, loan sharks, and agricultural college experts came forth with a panacea—ditching and draining. Swampland costing five dollars an acre could be cleared and drained for ten and then ought to be worth twenty-five, said the agricultural bulletin. Hundreds of thousands of acres in central Wisconsin were organized into districts and drained in the early years of the new century, and practically all of the projects failed.[9]

The stored fertility of the marshland was quickly exhausted.

Aldo Leopold's Sand Country

Rapidly declining crop yields left farmers saddled with debt, while the depressed water table left dry peat to be consumed by virtually inextinguishable fires. Leopold described such a burn: "Sun-energy out of the Pleistocene shrouded the countryside in acrid smoke. No man raised his voice against the waste, only his nose against the smell. After a dry summer not even the winter snows could extinguish the smoldering marsh. Great pockmarks were burned into field and meadow, the scars reaching down to the sands of the old lake, peat-covered these hundred centuries" (100). Fires ran at will over the sand counties during the 1920s, eating the heart out of abandoned lands. The worst fire year of all was 1930, when three hundred thousand acres of peat were consumed.

By 1933, when Leopold began working on land utilization problems with his new colleagues at the university, less than half the land in any of the sand counties was in farms, and of that very little was actively cultivated. The rest was considered waste-land—weeds, brush, runty jack pine, scrub oak, and raw peat sprouting dense thickets of seemingly worthless aspen. Much of the land had reverted to the counties for nonpayment of real estate and drainage taxes. There it reposed, for few would think of buying it.

It required the deepest economic depression in American history for people finally to confront, even half-blindly, the consequences of mindless expansion and exploitation, and to undertake, however falteringly, that process of social reorganization that Frederick Jackson Turner had called for two generations earlier in "The Problem of the West." When Turner suggested that the Old Northwest would be the battlefield on which the problems of American development would be resolved, he could hardly have dreamed that the very part of the region he knew as a boy would become one of the most blighted areas of the nation and an arena for experiment in social and institutional reorganization.

But within little over a year, from late 1933 to 1934, the course of development in the sand counties was dramatically reoriented. The rapid changes were turbulent, and to many the area must have seemed like Turner's witch's cauldron boiling with radical doctrines. Stirring the brew were administrators, technicians, and crews from a legion of new alphabetical agencies created by Franklin D. Roosevelt's New Deal—AAA, CCC, CWA, WPA, FERA, FSA, SES—as well as the old Biological Survey and Forest Service, various divisions of the Wisconsin Conservation Department and other state agencies, university departments, the agricultural extension service and county agents, county and town officials, and even a few local citizens. Overnight, projects were underway on classification and zoning of land; resettlement of families from submarginal farms to areas that might support productive agriculture and a self-sustaining community life; plugging of drainage ditches, construction of dikes, and reflooding of marshlands for wildlife, recreation, and cranberry growing; reforestation of sandy uplands; construction of firebreaks, lookout towers, and roads; and even some planting of game food patches and research on grouse. Out of all this came county forests, a state forest, several state and county parks (including one commemorating John Muir), the Necedah National Wildlife Refuge, the Central Wisconsin Conservation Area, numerous public hunting grounds, and state wildlife refuges, not to mention private ventures.

In all this ferment it is impossible to trace the precise influence of individuals like Leopold and his colleagues, though they were actively involved with most of the agencies, planning and negotiating, advising and criticizing. Their vision of the future of the sand country, including zoning and resettlement, reforestation and reflooding, was reflected in much of the feverish activity on the land. In his game survey of Wisconsin, Leopold had proposed state acquisition of some of the tax-reverted land in the sand counties for reforestation, reflooding, and management as combined public shooting grounds and public forests. Early in 1933, before the flurry of New Deal programs, he had proposed a Central Wisconsin Foundation to "pioneer beyond the usual and familiar categories of land use."[10] He hoped to find an economic use for the region's tax-reverted and idle lands, a wildlife crop that could coexist with the scattered farms and forest lands without major capital investment. The sand counties were unique in offering a variety of desirable game for every circumstance—wet or dry, forest or prairie, farm or wilderness. Along with the grouse moors of Scotland, which grossed five million dollars annually, they were a conspicuous exception to the general rule that the size and variety of the possible game crop varied directly with the agricultural value of land, and in that fact, Leopold argued, lay the special economic value of the sand counties as a wildlife area. Without doubt, the subsequent emphasis on the development of wildlife habitat in federal, state, and university projects can be attributed in large part to the far-reaching influence of Aldo Leopold.

Yet Leopold was profoundly dismayed by much of what he saw happening on the ground: clean-up crews taking out all the brush and hollow snags needed for wildlife food and shelter, road crews silting trout streams and gridironing the wilder expanses with unnecessary fire lanes, planting crews setting out jack pines in huge monotypic blocks. Integrated conservation had been accepted in theory ever since the days of Gifford Pinchot and Theodore Roosevelt, but it took the open money bags of 1933, Leopold wryly observed in a talk on "Conservation Economics," to reveal the ecological and esthetic limitations of "scientific technology," especially as practiced by singletrack agencies. Neither he nor any of the proponents of integrated conservation had ever before had enough field labor simultaneously at work on different projects to appreciate fully either the pitfalls or the possibilities. "If the *accouchement* of conservation in 1933 bore no other fruits," he concluded, "this sobering experience would alone be worth its pains and cost." But if trained technicians on public lands found it so difficult to integrate the diverse public interests in land use, what of the private landowner? [11]

The wholesale public expenditures of the New Deal indicated to Leopold that government might be persuaded to pay the bill for the ecological debt incurred by private exploitation and abuse of land. But government conservation, no matter how extensive and well administered, could not possibly go far enough. Wouldn't it make more sense to prevent environmental deterioration by encouraging good land use through demonstration, subsidy, and regulation, rather than to "cure" the abuses after the fact? At issue were two conflicting concepts of the desired end. One seemed to regard conservation as "a kind of sacrificial offering, made for us vicariously by bureaus, on lands nobody wants for other purposes, in propitiation for the atrocities which still prevail everywhere else." The other concept—Leopold's—supported the public program, especially in its teaching and demonstration aspects, but regarded government conservation only as an initial impetus, a means to an end. "The real end," he maintained, "is a *universal symbiosis with land,* economic and esthetic, public and private."

It was in this atmosphere, in the first flush of the New Deal, that Aldo Leopold first voiced his concept of a "conservation ethic." Noting the gradual evolution of ethics from individual to social relationships, he called for the extension of ethical criteria to the third element of the human environment, the land and the plants and animals that grow upon it.[12] The idea had evolved during his boyhood on the Mississippi, his years as a forester in the Southwest, and his early contact with the land use problems of central Wisconsin. But his full expression of a land ethic was the product of another order of experience, his own personal interaction with the land at his sand country shack.

Writing in later years, Aldo Leopold explained that he had bought himself a sand farm in an attempt to learn what it was about the sand counties that made destitute families unwilling to pull up stakes and resettle elsewhere, even when exhorted by government officials and baited with favorable credit terms. Though he eventually found an answer—and an equal attachment of his own to the sand country—his original acquisition of the place had not come about quite as he suggested.[13]

Among the conservation innovations of 1934 was Wisconsin's first bow-and-arrow deer hunting season. Leopold had been arguing for such a season for years, and his party of eight—four Leopolds and four friends—constituted twenty percent of the archers who took advantage of the opportunity that year. They hunted all five days of the season along river bottoms in the Southern sand counties, camping out in a tent; and though they failed to get a deer they saw plenty and had a marvelous time. It was enough to rekindle Leopold's old yen for a hunting camp. Ed Ochsner, one of the party, located an abandoned river-bottom farm and Leopold visited it with him for the first time on January 12, 1935.

The only building was a dilapidated chicken-house-turned-cowshed with manure knee deep on the floor. The farmhouse on the side of the hill had burned to the ground some years earlier, leaving only a fieldstone foundation. The island opposite the cowshed had been stripped of timber just a year or so before. The marsh across the road had apparently burned around 1930, the year of the great peat fires. But here the fires had not consumed the peat itself, for the river flooded the marsh each spring and the water table was too high. East of the shed was a corned-out field coming up to sand burs and panic grass, while up the hill to the west the sands were bare and blowing. But it would serve adequately for a hunting camp and it could be leased for a song. What's more, Ochsner reported, the owner would not mind at all if they wanted to do a little work around the yard or even build a house on the old foundation, so long as they didn't make a mess of the place. For Leopold's purposes the cowshed would suffice, but explaining it to his wife was a little delicate. All that manure.

Ten dollars for a lease, a "sort of" contract signed by the owner, and the Leopolds were ready to go to work: shoveling manure, building a fireplace, repairing the roof, battening the cracks in the walls. The place must have grown on them, for in early May Leopold wrote Ochsner that he had the notion to buy forty acres or so, provided it was obtainable at a proper price. By May 17 he was the owner of eighty acres of river-bottom land and a cowshed called "the shack." His first recorded act as a land-owner was to plant a food patch for wildlife, and his journal entries that summer, scant as they were, did not fail to mention the height of the sorghum.

The shack was a family enterprise to which each member contributed: cutting and splitting wood, building birdhouses for martins, screech owls, and wood ducks, planting prairie grasses and wildflowers, shrubs and trees. From April to October scarcely a day went by that someone did not plant or transplant something—butterfly weed, tamarack, wahoo and oak, june-grass and sideoats, penstemon and puccoon, pipsissewa and pasques. All five Leopold children pitched in. Starker built grape tangles for the birds, and for the family the one essential outbuilding which he dubbed the "Parthenon." Luna designed a new and better fireplace complete with a massive sandstone lintel and a handhewn cedar log mantel, rubbed to a mellow sheen. Nina and Carl were avid phenologists, observing and recording the annual order of events in nature—the first bloom of the pasques, the arrival of bluebirds, the fall of ripe acorns. In winter they banded resident birds, including 65290, the feisty little chickadee who regularly bloodied his beak in the trap but outlived all his fellows to be immortalized in *A Sand County Almanac*. Even Estella, youngest of the family, had her own projects, triumphantly constructing bridges to her secret island. Floods could be counted on to wash out her bridges each spring, insuring more days of happy engineering.

For Leopold and his family the shack years were an experience in the slow sensitizing of people to land, the evolution of a sense of country. The shack originally acquired as a hunting camp soon became a "weekend refuge from too much modernity" (viii), a place to hike and swim and savor the outdoors, to build with their own hands, to split oak and make sourdoughs in the dutch oven at an open fire, to play guitars and sing and talk and laugh together. It was also a place where one could experience a feeling of isolation in nature—especially if one were an insomniac like Leopold, who habitually arrived "too early" in the marsh. And it offered rich country for the growth of perception. The more woodcock nests they discovered, the more trees and shrubs, grasses and flowers they planted, the more chickadees and nuthatches they got to know—in short, the more familiar they became with the place—the more they found to anticipate, to ponder, and to marvel at. The journals of their shack visits reveal an almost exponential increase in recorded observations over the years, and *A Sand County Almanac* is eloquent testimony to the meaning and value of the experience.

Most important, the shack offered space enough and time to practice the arts of wild husbandry. A sense of husbandry, said

Leopold, "is realized only when some art of management is applied to land by some person of perception" (175). Wild husbandry offered a substitute for what he termed the "split-rail value" of the pioneer tradition, symbolized by Daniel Boone and characterized by free-for-all exploitation of land. Like the split-rail tradition, it was a reminder of the elemental man-earth relation, but in addition wild husbandry required ethical restraint in the use of tools, and thus had special cultural value for mechanized man.

Nothing better illustrates Leopold's sense of husbandry than his use of shovel and axe in planting and encouraging his pines. Two thousand, three thousand, sometimes five or six thousand pines a year, every year, Leopold and his family planted in an annual weeklong ritual, the spring planting trip. They planted them all by hand with shovels so sharp they sang and hummed in their wrists as they sliced the earth. And they planted them with care, in groves, points, and stringers, or interspersed with other trees and shrubs.

But not all of the pines thrived. The first year, 1936, 95 percent of the Norways and 99 percent of the whites were killed by the drought within three months of planting. The next summer drought losses were heavy again, even among hand-watered seedlings. Leopold's journals contain cryptic hints of his tribulations: "Never plant pines near grape or poison ivy," or "You can't put up brush shelters"—which attract rabbits—"and plant white pine at the same time and place." In one winter rabbits trimmed three-quarters of the white pines planted in the woods. Another winter, deer were the culprits, and, in the spring, floods drowned out seedlings on lower ground. To compensate, Leopold planted more pines more thickly. He planted pines in grass or beneath nurse trees, in plowed ground or in furrows. He tried spoon-feeding them fertilizer and mulching with marsh hay. He tried weeding by hand, and surrounding each tree with cardboard to kill the grass, and "desodding"—scalping an area of sod around each tree. He became so fond of the desodding technique that the family figured the time it took him to walk home through his pines, scalping all the way, was equal to the square of the distance. But after all this care he could still lose pines to drought or flood or rabbits or deer or rust or weevils, or get bad planting stock from the nursery to begin with, or have birds alight on the candles and break them off, or vandals cut off the leaders out of sheer meanness. Nevertheless, some of the pines managed to survive, and within three years Leopold proudly measured them against little Estella—up to her collar or her nose or the top of her head.

Then came the fires. In November 1941 a campfire got away from a trespassing hunter and burned an area of pines near the shack before it was put out by several of the neighbors. The following March a much larger fire of more mysterious origin burned most of the marsh and all but a few of the pine plantings. All that summer the Leopolds watched their pines dying. Many had been killed outright. Others that looked healthy one week were dead the next. An infestation of powder-post beetles spread from fire-killed trees to wounded survivors, so the Leopolds cut and burned all dead and dying trees. Still the toll mounted. Yet fire brought life as well as death. Willows and aspens, wild plum, sumac, and hazel resprouted vigorously; dewberry and blackberry, bluestem and lespedeza increased, and so did ragweed and poison ivy. The next summer Leopold was thrilled to find his first natural reproduction—four young jackpines eight inches high growing in a scalp beside a dead white pine, undoubtedly the legacy of cones opened by fire.

So it went, the husbandry of pines—the painstaking care in planting, the anxious days of weeding, watering, watching, the toll of drought, flood, and fire, rabbit and weevil, the pride in measuring the growth of trees that thrived. There was a special dividend from pines that had overcome such adversity. In the snowy stillness of a midwinter evening when "the hush of elemental sadness lies heavy upon every living thing," there were Leopold's pines standing ramrod straight under their burden of snow, and in the dusk beyond he could sense the presence of hundreds more. "At such times," he wrote, "I feel a curious transfusion of courage" (87).

If the shovel symbolized the planting and care of struggling young seedlings, the axe characterized decisions required of the husbandman as his vigorous saplings raced skyward. The 1940s

Aldo Leopold's Sand Country

were years of the axe as well as the shovel at the Leopold shack, and the axe-in-hand decisions seemed somehow more difficult, perhaps because of their finality. Where a birch was shading a pine, he usually cut the birch. But what of a veteran oak with pines heading toward its outthrust limbs? Or a cluster of pines planted close in the hope that one would survive—had they all earned the right to compete for greater glory? A wilderness purist might let the trees fight it out among themselves. But the shack was not pure wilderness, nor was Leopold merely a spectator. The land had been heedlessly ravaged by men who regarded it as a commodity to be used and then abandoned. Leopold, by contrast, regarded himself as a participating citizen of the land community, seeking to restore it to ecological integrity, and he would not shirk the ethical decisions this entailed.

Through a lifetime of observation and experience, of perception and husbandry, Aldo Leopold had clarified his understanding of ecological processes and the fundamental values—integrity, stability, and beauty—that he saw as the basis of a land ethic. But ethical values were a guide for individual decisions, not a substitute for them, and Leopold realized this most keenly when he stood with axe in hand. Conservation, he wrote, "is a matter of what a man thinks about while chopping, or while deciding what to chop. A conservationist is one who is humbly aware that with each stroke he is writing his signature on the face of his land." And then an allowance for the subjective hopes, ideals, affections, and convictions of the individual: "Signatures of course differ, whether written with axe or pen, and this is as it should be" (68).

Perception honed in the practice of husbandry engendered in Leopold a profound humility in his use of tools. Through his own decisive participation in the land community he became acutely aware of the innumerable, ofttimes inscrutable factors involved in life and death, growth and decay. Not all pines thrived even with the greatest foresight and care, nor did all industrious pioneers turn a profit on their land, nor did all scientific technologies or all alphabetical agencies inevitably yield progress. Though it took some of the certitude out of individual decisions and individual existence, the practice of husbandry gave Leopold a sense of belonging to something greater than himself, a continuity with all life through time. This intellectual humility rooted in perception led Leopold to appreciate that "all history consists of successive excursions from a single starting-point, to which man returns again and again to organize yet another search for a durable scale of values" (200).

For Aldo Leopold the sand country was such a starting point. It was a backwater refuge from the heedless rush of progress, a setting in geological time that he shared with the sandhill cranes, the pines, and the pasques of lineage more ancient than man. At the shack, through his unique capacity for perception and husbandry, he became a participant in the drama of the land's workings, and he transformed the land as it transformed him.

Both the surface and the depths of Leopold's shack experience he expressed in the essays that now comprise *A Sand County Almanac*. At the urging of his closest friends Leopold began as early as 1941 to seek a publisher for his essays in book form. A succession of publishers turned him down over the years. The essays lacked cohesion, they said, or their ecological and philosophical concepts were too difficult for laymen, or the book would not find a market to warrant using scant wartime allotments of paper. Though undoubtedly discouraged, Leopold continued to write essays, and late in 1947 he sent a much expanded and thoroughly restructured version of the manuscript to still other publishers. By this time he was committed to bringing the book out himself if he had to. But on Wednesday, April 14, 1948, Oxford University Press notified him by long-distance telephone that they would be delighted to publish the book.

Leopold left for his annual spring planting trip two days later a happy man. With him were his wife and his daughter Estella, a senior at the University of Wisconsin and the only one of the five children still in Madison. All shack trips were enchanted, but this time everyone was in especially good spirits. Pasques were blooming, tamarack buds bursting, and chickadees were getting paired up, though one banded chick was still without a mate. Ruffed grouse drummed until dark, woodcocks peented at dusk and continued by moonlight, accompanied by goose music on the marsh. On Monday Leopold picked up the pines, only two hundred whites and two hundred reds, fewer than ever before. He was still recuperating from an operation for tic douloureux and was under heavy strain from an overload of students, so he intended to take it easy that year.

That evening, as the geese started to come back from corn, the three Leopolds dressed in khaki brush-colored clothes and walked out to a bench on one of the aspen islands in the marsh to watch them come in. Just as the sun hit the horizon, geese began arriving in bunches, coasting in low over their heads or "maple leafing" from a great height, tilting first one wing low and then the other to lose altitude. Leopold with his stubby pencil and little black notebook jotted down all the flock counts—5,5,9,2,3,2,8 and so on to a total of 445. He was trying to check the hypothesis that geese flew in family groups. Yet it was not scientific research that drew the Leopolds into the marsh so much as the pure enjoyment of the performance. A basic human quality was the ability to appreciate simple things that could not be eaten or worn or sold, Leopold had told his daughter while they were sweeping the shack floor one morning, and now as they sat on the bench in the marsh and exchanged glances and nudged each other in their enthusiasm over each new flock that came in she knew what he meant. The geese were "a last bit of real wild life in our time, probably the only quantity of game left in these parts," Leopold said with a touch of nostalgia while he entered the day's notes in the journal back at the shack that evening, as his wife knitted and Estella sat on the doorstep playing the guitar and singing softly.[14]

The next morning Leopold was up by 4:30 A.M. as usual, listening for the daybreak songs of the birds and recording the times and light intensities in his notebook while it was still so dark he had to shape the numbers by feel. Later that morning he planted about a hundred pines with Estella, cleared a trail through the prickly ash to the otter pool, and cut some sprouts in the birch row. A pileated woodpecker flew past four times. In the afternoon he and his wife went down to inspect Estella's new bridge, then paddled slowly up the slough in the canoe, putting up wood ducks and bluewings. He went straight to bed after writing his journal entry that evening, and Mrs. Leopold confided to Estella her worry about his tiredness.

Wednesday, April 21, dawned clear, calm, and cold at the shack. That morning Leopold counted geese heading outward to corn between 5:15 and 5:40 A.M. His total was 871, more geese than he had ever before counted at the shack. "On our farm we measure the amplitude of our spring by two yardsticks: the number of pines planted, and the number of geese that stop," he had written in his essay "The Geese Return," which he had revised for the book just a month earlier. "Our record is 642 geese counted in on 11 April 1946" (19). April 21, 1948, was a new record, though he did not make note of the fact when he transferred his flock counts to the journal that morning.

The three Leopolds were in high spirits as they busied themselves around the shack after breakfast, repairing all the broken tools which Leopold termed "a disgrace to the outfit." Around 10:30 they spotted smoke coming from the east across the marsh on a light breeze. "Someone's burning his hay meadow," Leopold commented as he continued with his tools. Then suddenly he became very concerned and excited, sent Estella after the Indian fire pump in the shack, and tossed buckets and a sprinkling can, coats, gloves, and brooms in the car. The three drove east along the birch row a half-mile to his new neighbor's farm, where they found a large area already burned and flames in places leaping five to eight feet high. About ten neighbors were there already fighting the blaze with pumps and buckets but it was out of control, moving rapidly toward the goose marsh and the pines beyond. Leopold told Estella to drive to the nearest phone and call the fire department, stationed his wife at the road with a broom to keep the fire from crossing, grabbed the sprinkling can, and headed toward the flames.

"Professor Aldo Leopold, Burned Fighting Grass Blaze, Dies" said the headline in the Madison paper that evening. But Aldo Leopold had fought too many fires in his life to be downed by flames. It was a coronary attack, the doctors decided when they heard the circumstances. He had fallen on unburned grass, probably not long after he had left his wife and daughter, and the fire had swept lightly over him sometime later.

Aldo Leopold's Sand Country

Aldo Leopold's passing left a void in his family, among his friends, and in the world of conservation thought. But it was a void ultimately filled by the imperishable force of his spirit and by the little book that embodied it. The book contains no panaceas, no blueprints for mass action. It is simply one man's expression of his experience with the land, offered to others who would search in their own way, in their own time and place, for the larger meaning and purpose in life. *A Sand County Almanac* was published in 1949 with a new foreword Leopold had written just the month before his death. "There are some who can live without wild things," he had begun, "and some who cannot. These essays are the delights and dilemmas of one who cannot" (vii).

Notes

1. The geologic history is based on Lawrence Martin, *The Physical Geography of Wisconsin*, 3d ed. (Madison: University of Wisconsin Press, 1965); and H. E. Wright, Jr., and David G. Frey, eds., *The Quarternary of the United States* (Princeton: Princeton University Press, 1965); and other sources.

2. For vegetation change see John T. Curtis, *The Vegetation of Wisconsin* (Madison: University of Wisconsin Press, 1959); and Wright and Frey, *Quarternary*.

3. For Indian history see George I. Quimby, *Indian Life in the Upper Great Lakes* (Chicago: University of Chicago Press, 1960); and Harold Hickerson, *The Chippewa and Their Neighbors* (New York: Holt, Rinehart and Winston, 1970).

4. John Muir, *The Story of My Boyhood and Youth,* (Boston: Houghton Mifflin Co., 1917), 51–52, 53, 59, 60, 180.

5. Ibid., 177, 176.

6. Ibid., 180–81.

7. Frederick Jackson Turner, *The Frontier in American History* (New York: H. Holt and Co., 1920), 2. For Turner's years in Portage, see Ray A. Billington, "Young Fred Turner," *Wisconsin Magazine of History* 46 (Autumn 1962): 38–48.

8. Turner, *Frontier,* 220–21.

9. See A. R. Whitson and E. R. Jones, *Drainage Conditions of Wisconsin,* Bulletin 146 (Wisconsin Agriculture Experiment Station, 1907).

10. Aldo Leopold, "A Proposed Central Wisconsin Foundation," typescript, c. 1933, p. 3, Leopold Papers (LP) 3B1, University of Wisconsin–Madison Archives.

11. Aldo Leopold, "Conservation Economics," *Journal of Forestry* 32 (1934): 537–44.

12. Aldo Leopold, "The Conservation Ethic," *Journal of Forestry* 31 (1933): 634–43.

13. The following account of Leopold's shack experience is largely based on the "Leopold Shack Journal," 3 vols. holograph, 1935–1948, LP 7B3; correspondence in the Baraboo Cabin file, LP 8B1; *A Sand County Almanac;* and personal interviews.

14. Estella Leopold to Nina Elder, April 1948, LP 8B9.

3 * Aldo Leopold's Intellectual Heritage

RODERICK NASH

The Greek word *oikos,* meaning house, is the root of both *economics* and *ecology.* Over time the significance shifted from the house itself to what it contained—a living community, the household. Economics, the older of the two concepts, concerns the study of how the community manages its material resources. Ecology took shape in the 1890s as the science of organisms (not just human ones) and how they interact with each other and with their environment. Ecology, in a word, concerned communities, systems, wholes.[1] And because of this holistic orientation, the ecological perspective proved to be fertile soil for environmental ethics. Aldo Leopold, himself a pioneer American ecologist, made the connection explicit. "All ethics," he wrote in *A Sand County Almanac,* "rest upon a single premise: that the individual is a member of a community of interdependent parts. . . . The land ethic simply enlarges the boundaries of the community to include soils, waters, plants, and animals, or collectively: the land" (203).

Steeped as their origins were in the social contract theory of John Locke, Americans took the concept of community to be replete with strong ethical overtones. Once an American conceded that something was a member of his *community,* the argu-

This essay draws upon several chapters in the author's forthcoming book, *Widening the Circle: Ethical Extension and the New Environmentalism.*

ment for its natural rights, inalienable since derived from the pre-government state of nature, was difficult to deny. As the definition of community broadened, so did the circle of ethical relevancy. In the century after the Revolution, American ethics underwent significant expansion. The abolition of slavery in 1865 was the most dramatic instance. In subsequent years the rights of women, Indians, laborers, and free blacks received attention. One of the most radical frontiers of ethical expansion was the human relationship to that other component of the state of nature: nature itself. When ecology helped Americans think of other species and the biophysical world as an oppressed and exploited minority within the extended moral community, the contemporary environmental movement received its most characteristic insignia. Old-style conservation, plugged into American liberalism, became the new environmentalism. Leopold's land ethic was one of its philosophical cornerstones because it conceived of the protection of nature in moral, not just the traditional economic, terms.

The expansion of natural rights to include the rights of nature depended on the realization that the community to which human beings belonged did begin and not end with other humans. Even before ecology provided scientific grounding for this assumption, there were anticipations of the idea in American thought. So while Aldo Leopold was a major figure in the development of both ecological thought and moral philosophy, his originality must not be distorted. Little, after all, is new under the sun, and in the history of ideas the adage rings especially true. Whether he acknowledged it or not, Leopold's achievements rested on more than a century of theological, philosophical, and scientific thought.

Henry David Thoreau's moral environmentalism stemmed from his Transcendentalist belief in the existence of an "Oversoul" or a godlike moral force that permeated everything in nature.[2] Thoreau's organicism, reinforced by both science and religion, led him to remarkable ideas about his environment. He repeatedly referred to nature and its creatures as his *society*, transcending the usual human connotation of that term. "I do not," he wrote in his journal for 1857, "consider the other animals brutes in the common sense."[3] He regarded sunfish, plants, skunks, and even stars as fellows and neighbors—parts of his community. "The woods," he declared during an 1857 camping trip in Maine, "were not tenantless, but choke-full of honest spirits as good as myself any day."[4] There was no hierarchy or any discrimination in Thoreau's concept of community. "What we call wildness," he wrote in 1859, "is a civilization other than our own."[5] No American had ever thought of anything faintly resembling this before.

Although he did not use the term *ethics,* an environmental ethic sprang from Thoreau's expanded community consciousness. It began with the axiom that "every creature is better alive than dead, men and moose and pine trees," and went on to question the almost omnipresent belief of his era in the appropriateness of human domination (kindly or not) over nature.[6] "There is no place for man-worship," he declared in 1852; and later: "The poet says the proper study of mankind is man. I say, study to forget all that; take wider views of the universe."[7] Thoreau's own expanded vision led him to rant against the Concord farmers who engaged in the quintessentially American activity of clearing the land of trees and underbrush: "If some are prosecuted for abusing children, others deserve to be prosecuted for maltreating the face of nature committed to their care."[8] Here Thoreau seemed to be implying that nature should have legal rights, as did powerless people subject to oppression. Making the same connection on another occasion, he pointed out the inconsistency of the president of an antislavery society wearing a beaver-skin coat. While Thoreau avoided the word *rights,* as well as *ethics,* his association of abused nature with abused people charted a course that the new environmentalists would later follow. But in the middle of the nineteenth century, Thoreau was not only unprecedented in giving voice to these ideas, he was virtually alone in holding them.

George Perkins Marsh's 1984 book, *Man and Nature: Or, Physical Geography as Modified by Human Action,* was the first comprehensive description in the English language of the destructive impact of the march of human civilization on the environment. In the manner of subsequent ecologists, Marsh wrote about the "balances" and "harmonies" of nature, but, unlike Thoreau, he did not move on to challenge anthropocentrism. He was perfectly content with the idea of mankind's dominion over nature, provided that it was careful and farsighted. That it had not been so was the subject of his book. "Man has forgotten," Marsh thundered, "that the earth was given to him for usufruct alone, not for consumption, still less for profligate waste." Anticipating the ecological perspectives of the twentieth century, Marsh warned that the interrelatedness of "animal and vegetable life is too complicated a problem for human intelligence to solve, and we can never know how wide a circle of disturbance we produce in the harmonies of nature when we throw the smallest pebble into the ocean of organic life." As a corrective to previous human carelessness, Marsh proposed "geographical regeneration," a great healing of the planet beginning with the control of technology. And this, he continued, would require "great political and moral revolutions."[9]

For the present purposes, what is most significant about Marsh's enormously influential book is that it marks the first published American discussion of conservation in ethical terms. Granted, Marsh's work contains nothing about the rights of nature; the welfare of people is constantly uppermost in his mind. But he does suggest that human custodianship of the planet is an ethical or "moral" issue, not just an economic one. It is right, in other words, to take care of nature; wrong to abuse her.

For John Muir, as for Thoreau and the earlier organicists, the basis of respect for nature was the perception of it as part of a created community to which man, also, belonged. God permeated Muir's environment. Not only animals, but plants and even rocks (Muir wrote about "crystals") and water were "sparks of the Divine Soul."[10] But civilization, and particularly Christianity with its dualistic separation of people and nature, obscured this truth. To reemphasize it, Muir deliberately chose to defend organisms at the bottom of the Christian's hierarchic chain of being— like snakes. "What good are rattlesnakes for?" he asked rhetorically. And, answering his own question, replied that they were "good for themselves, and we need not begrudge them their share of life."[11]

Aldo Leopold's Intellectual Heritage

Muir made a related point in reference to the alligators he encountered on an 1867 hike through Florida. Commonly regarded as ugly and hateful vermin, he preferred to understand the giant reptiles as "fellow mortals" filling the "place assigned them by the great Creator of us all" and "beautiful in the eyes of God." Later, in the same 1867 journal, Muir drew the full implications of his novel ideas: "How narrow we selfish, conceited creatures are in our sympathies! How blind to the rights of all the rest of creation!" Here, in Muir's journal, was the first association of "rights" with nature in American intellectual history. Its basis lay in Muir's perception of people as members of the natural community: "Why," he asked on the 1867 walk, "should man value himself as more than a small part of the one great unit of creation?"[12] He also gave unprecedented American articulation to the organicists' vision: "When we try to pick out anything by itself, we find it hitched to everything else in the universe."[13]

Darwinism took the conceit out of man. *On the Origin of Species* (1859) and particularly *The Descent of Man* (1871) were important conceptual resources for Aldo Leopold's eventual environmental ethic. The evolutionary explanation of the proliferation of life on earth undermined dualistic philosophies at least two thousand years old. Charles Darwin put man back into nature. He broadened the meaning of kinship. No more special creation in the image of God, no more immortal "soul," and, it followed, no more dominion or expectation that the rest of nature existed to serve one precocious primate.

Darwin himself was well aware of the ethical implications of his evolutionary hypothesis. As early as 1837 he referred to animals as "our fellow brethren" and remarked that "we may be all melted together."[14] True, Darwin described fierce competition, but he saw a commonality among all the competitors. Living and dying together over the eons, everything alive participated in a universal kinship or brotherhood. The idea of respect for man's fellow participants followed immediately, the expression of this respect being one of the distinguishing marks of civilized people. In 1871, in *The Descent of Man,* Darwin addressed the matter directly. Chapter 4 of the book discusses the idea that the moral sense (or, as Darwin preferred, "sympathy") was a product of evolution just like the eye or the hand. He believed ethics had arisen from a preethical condition where self-interest alone existed. Over time humans broadened their ethical circle to include "small tribes" then "larger communities" and eventually "nations" and "races." It reached out still further "to the imbecile, maimed, and other useless members of society." And then Darwin made an extraordinary conceptual leap: "Sympathy beyond the confines of man . . . to the lower animals, seems to be one of the latest moral acquisitions." Ultimately, Darwin thought that, as ethics evolved, all "sentient beings" would come to be included in the moral community.[15]

In shaping his view on the evolution of ethics, Darwin drew, as his footnotes in *The Descent of Man* indicate, on the work of William E. H. Lecky. This distinguished Irish intellectual historian published his *History of European Morals from Augustus to Charlemagne* in 1869 at a time calculated to influence Darwin who was then in the final stages of writing *The Descent*. Lecky endeavored to show that "there is such a thing as a natural history of morals, a defined and regular order, in which our feelings are unfolded." As Lecky saw it, this process had improved moral standards since Roman times. The recent inclusion of animals in the ethical circle was an important part of Lecky's evidence for this theory. Darwin's footnote states that Lecky "seems to a certain extent to coincide" with his own conclusions.[16] In point of fact Lecky anticipated Darwin almost completely.

While slower-developing than its counterpart in England, the American humane movement, and its components such as vegetarianism and antivivisectionism, deserve more recognition than they have received as intellectual precursors of environmental ethics. One contemporary environmental philosopher, J. Baird Callicott, goes so far as to argue that, in their shallow and limited perspective, the animal liberationists are not even fellow travelers with the deep ecologists and new environmentalists. He quotes an advocate of holistic environmental ethics to the effect that "the last thing we need is simply another 'liberation movement.'"[17]

Aldo Leopold's Intellectual Heritage

It is impossible, of course, to deny that nineteenth-century anticruelty efforts in the United States were largely confined to domesticated animals. Pain in familiar creatures was the evil to be eradicated. But Callicott's somewhat ahistorical contemporary criticism does an injustice to the old humanitarians. Granted, they lacked philosophical consistency as well as a holistic ecological consciousness. If pain was the problem, as Mark Sagoff has argued, the animal liberationists should attempt to end it in wild as well as domestic contexts. Why not, Sagoff asks tongue in cheek, try to feed and shelter the millions of wild creatures who die horrible deaths in the wilderness every day?[18] But this satire is too abrupt. Clearly the nineteenth-century humanitarians were not biocentric; they did not consider natural systems so much as individual organisms (and only certain of those) in moral terms. But they do deserve credit for making the first stumbling steps away from an ethic (or moral community) that began and ended with human beings. As Lecky and Darwin understood, there was a historical progression in the evolution of morality that began closest to home. Animals were the next in line following people. So it is not particularly helpful to fault the humanitarians, then or now, for not going further with ethical extension. What they did do was revolutionary enough in terms of the main currents of Western ethical thought.

The most direct American anticipations of Aldo Leopold's ethical ideas occur in the writings of Edward Payson Evans and J. Howard Moore. Evans wrote books about animal symbolism in architecture and about the prosecution of animals in ancient and medieval courts of justice. The first product of Evans' research into ethics and psychology was an article in the *Popular Science Monthly* for September 1894 entitled "Ethical Relations Between Man and Beast." It began with a statement of Evans' intent to correct the "anthropocentric assumption" in psychology and ethics just as it had been corrected over the past several centuries in astronomy and, more recently, by Darwin in biology. Evans' initial task was to attack the religious basis of anthropocentrism. Anticipating Lynn White's 1967 thesis by seventy years, Evans developed a remarkably full case against "the anthropocentric character of Christianity." Evans criticized the "tyrannical mandate" contained in Genesis to conquer the earth, and he compared Judeo-Christianity unfavorably to the biocentric religions of the East such as Buddhism and Brahmanism.[19] Taken for granted by environmentalists after the 1960s, this exposure of the shortcomings of Western religious tradition was unanticipated in the 1890s even in the writings of so sharp a critic of orthodox Christianity as John Muir.

In the 1894 essay, but more extensively in a book three years later entitled *Evolutional Ethics and Animal Psychology*, Evans proceeded to demonstrate the commonality of humans and animals on "strictly scientific grounds." Drawing on evolutionary biology and on the new field of animal psychology, he discussed "metempsychosis" or reincarnation, communication abilities, the aesthetic sense, and the universality of "consciousness" or what he calls "mind." In these matters Evans often skated boldly over ice that later psychologists and philosophers would find thin. But his real purpose was to exhort, and this he accomplished with unprecedented enthusiasm: "Man is as truly a part and product of Nature as any other animal, and [the] attempt to set him up on an isolated point outside of it is philosophically false and morally pernicious." On this basis Evans branded as wrong "maliciously breaking a crystal, defacing a gem, girdling a tree, crushing a flower, painting flaming advertisements on rocks, and worrying and torturing animals."[20] For Evans, nonhuman life forms had intrinsic rights that man must not violate.

The stated topic of his 1894 book was the *evolution* of ethics, and Evans devoted considerable attention to the changes in morality over time. He believed he was riding the crest of an intellectual wave. While Evans' indebtedness to Charles Darwin and William Lecky is obvious, in one respect Evans' work is unprecedented. Even the most vigorous of the nineteenth-century humanitarians and natural rightists stopped with animals. Evans, however, goes far beyond, apparently to every living thing and even, as observed, to inanimate objects like rocks and minerals.

J. Howard Moore's major statements appeared in 1906 and 1907 under the titles *The Universal Kinship* and *The New Ethics*. Henry Salt, the leading English humanitarian, published the

first book, undoubtedly with the hope of closing the gap between America's nascent animal rights movement and England's
well-established one.

Along with Darwinism, misanthropy was the driving force
behind Howard Moore's ethical philosophy. A man, in his eyes,
was "not a fallen god, but a promoted reptile." In fact, men were
worse than snakes: "the most unchaste, the most drunken, the
most selfish and conceited, the most miserly, the most hypo-
critical, and the most bloodthirsty of terrestrial creatures."
Gradually and incompletely Moore felt, people have tried to
transcend their innate selfishness but they were still mired in
what he termed the "larval stage." The extension of ethics beyond humans was a hopeful sign. Moore, in fact, understood his
"new ethics" to be the cutting edge of what he grandiloquently
styled "the great task of reforming the universe."[21]

The starting point for Moore's ethical system, as his title implies, was universal kinship. To Moore this meant that "all the
inhabitants of the planet Earth" were related "physically, men-
tally, morally." He took pains to explain that his sense of ethical
community applied "not to creatures of your own anatomy only,
but to *all* creatures." It followed for Moore that the Golden
Rule, or what he termed the "Great Law," was "a law not applicable to Aryans only, but to *all* men; and not to men only, but to
all beings." Moore thought of this principle as "simply the expansion of ethics to suit the biological revelations of Charles
Darwin."[22]

At various points in his two books Moore adopted a far-
reaching ethical stance. "All beings are *ends,*" he preached. "*No*
creatures are *means.*" And again: "All beings have not equal
rights, but *all have rights.*" But it soon becomes clear that Moore's
ethical circle has definite limits. He respected the rights of domestic animals, of course, and also of mice, turtles, insects, and
fish. But plants have no status in Moore's moral system. In *The
New Ethics* he explained that plants lacked "*consciousness*"—they
were alive but did not "*feel.*" Consequently, they could neither be
harmed nor their rights violated. Plants were "outsiders," "mere
things," and not beings or creatures.[23]

It is likely that Moore's ardent vegetarianism contributed to this attitude. But it is also pertinent that in the 1910s he would not have had the benefit of the ecologists' sense of natural process, which did not attach a moral status to participation in a food chain. As it is, Moore strikes contemporary ecologists as rather ridiculous, arguing, for instance, against meat eating even by wild animals. Indeed ecologists are almost unanimous in deprecating the animal liberation/rights argument. Nonetheless, much of what Moore wrote is impressive even by the standards of latter-day ethically minded ecologists such as Aldo Leopold, a college student when Moore's books appeared. For example, Moore wrote about "earth-life as a single process . . . every part related and akin to every other part." And he could say "the *Life Process is the End—not man.*"[24] Clearly, Moore's ideas, like those of Edward Evans, were stepping stones toward a more comprehensive environmental ethic.

Good evolutionists as they were, both Evans and Moore saw themselves as part of an unfolding intellectual process. Their own age might be ethically unenlightened, but they could be optimistic about the future. Charles Darwin, their mentor, had been confident about the spread of altruism through natural selection. The relatively recent abolition of slavery and gains in rights for laborers and women encouraged them. Moore made the links explicit: "The same spirit of sympathy and fraternity that broke the black man's manacles and is to-day melting the white woman's chains will to-morrow emancipate the working man and the ox." Moore understood that this emancipation would not occur overnight. "New ideas," he wrote in *The New Ethics,* "make their way into the world by generations of elbowing." But despite the lack of respect for his opinions among his contemporaries, Moore could look forward to a time when "the sentiments of these pages will not be hailed by two or three, and ridiculed or ignored by the rest; *they will represent Public Opinion and Law.*"[25] Evans also forecast a day when "our children's children may finally learn that there are inalienable animal as well as human rights."[26] His use of natural rights theory and even Jeffersonian rhetoric showed the ease of transferring those concepts from humanity to nature. And his sense of timing was uncanny. The "children's children" of Evans' generation lived after 1960,

Aldo Leopold's Intellectual Heritage

and some of them celebrated Aldo Leopold and became deep ecologists and new environmentalists.

The Cornell University horticulturist, Liberty Hyde Bailey, concluded three decades of what he termed a "biocentric" approach to his subject with the publication, in 1915, of *The Holy Earth*. The book followed a well-worn intellectual path in arguing that abuse of the earth was morally wrong because it was God's creation. Bailey went on to advocate abandoning "cosmic selfishness" and developing a sense of "earth righteousness." This, he felt, would help human beings "put our dominion into the realm of morals."[27]

Probably because he couched his philosophy in simple terms and lived it dramatically in the heart of Africa, Albert Schweitzer had a much greater impact on the development of environmental ethics in the United States than Bailey (though Bailey, as a fellow American natural historian, was almost certainly better known to Aldo Leopold). In September 1915, while on a small steamer moving up the Ogowe River, Schweitzer discovered in the phrase "Reverence for Life" the most valid basis for ethics. He built a theory of value based on a concept evidently borrowed from the philosophy of Arthur Schopenhauer, the "will-to-live," which every living being possessed. Right conduct for a human consisted of giving "to every will-to-live the same reverence for life that he gives to his own."[28]

Schweitzer made it abundantly clear that his reverence for life did not end with human beings. In fact he commented that "the great fault of all ethics hitherto has been that they believed themselves to have to deal only with the relations of man to man." In his eyes "a man is ethical only when life, as such, is sacred to him, that of plants and animals as that of his fellow men."[29] In other writings Schweitzer went still further, apparently extending his ethics to all matter. The ethical person, Schweitzer wrote in 1923, "shatters no ice crystal that sparkles in the sun, tears no leaf from its trees, breaks off no flower, and is careful not to crush any insect as he walks."[30] So Schweitzer would place a worm, washed onto pavement by a rainstorm, back into the grass; he would remove an insect struggling in a pool of water. The powerful and privileged status humans enjoyed in the natural community entailed for Schweitzer not a right to exploit but a responsibility to protect.

Like William Lecky and Charles Darwin, Schweitzer concerned himself with the history and future of ethics. He believed in the potential of ethical evolution. He wrote that the thoughtful person must "widen the circle from the narrowest limits of the family first to include the clan, then the tribe, then the nation and finally all mankind." But this was only the beginning for Schweitzer. "By reason of the quite universal idea . . . of participation in a common nature, [one] is compelled to declare the unity of mankind with all created beings." Of course Schweitzer understood that so fundamental an intellectual revolution as the extension of ethics to new categories of beings was not easy. And World War I reminded him of the shortfall in even person-to-person ethics. But he took hope from the history of ideas: "It was once considered stupid to think that colored men were really human and must be treated humanely. This stupidity has become a truth." In the same manner Schweitzer predicted in 1923 that the circle would continue to widen: "Today it is thought an exaggeration to state that a reasonable ethic demands constant consideration for all living things down to the lowliest manifestations of life. The time is coming, however, when people will be amazed that it took so long for mankind to recognize that thoughtless injury to life was incompatible with ethics." [31]

Albert Schweitzer's ideas reached the United States in English translations of his books in the 1920s and 1930s. Although his was a mystical holism, it coincided remarkably with the ecologists' concept of a biotic community. No life was worthless or merely instrumental to another life; every being had a place in the ecosystem and, some philosophers and scientists were beginning to think, a right to that place.

Few today would challenge Aldo Leopold's reputation as one of the seminal thinkers in the modern American development of environmental ethics. Yet his statement of "the land ethic," the basis of his enormous reputation, amounts to but twenty-five undocumented pages at the conclusion of a book he did not live to see in print, *A Sand County Almanac*. Nevertheless, within

two decades Leopold's manifesto became the intellectual touch-
stone for the most far-reaching environmental movement in
American history. In 1963, Secretary of the Interior Stewart L.
Udall declared that "if asked to select a single volume which contains a noble elegy for the American earth and a plea for a new
land ethic, most of us at Interior would vote for Aldo Leopold's
A Sand County Almanac." J. Baird Callicott called Leopold "the
father or founding genius of recent environmental ethics," a
writer who created the standard or "paradigm" of an ethical system that included the whole of nature and nature as a whole.
Wallace Stegner thought *A Sand County Almanac* was "one of
the prophetic books, the utterance of an American Isaiah," and
Donald Fleming, Harvard's historian of ideas, called Leopold
"the Moses of the New Conservation impulse of the 1960s and
1970s, who handed down the Tablets of the Law but did not live
to enter the promised land." In a similar vein, Clay Schoenfeld
singled out Leopold as "an authentic patron saint of the modern
environmental movement, and *A Sand County Almanac* is one of
its new testament gospels." Van Rensselaer Potter dedicated a
1971 volume to Leopold as one "who anticipated the extension of
ethics to Bioethics." *A Sand County Almanac,* which Leopold
thought might never even find a publisher, went on after his
death to sell a million copies in several paperback editions.[32]

Leopold graduated in 1909 from Yale University's School of
Forestry, which owed its existence to the generosity of Gifford
Pinchot, and understandably absorbed much of the utilitarianism
of Pinchot and the progressive conservationists. Nature was to
be used—albeit wisely and efficiently—for the greatest good of
the greatest number (of people, of course) in the longest possible
run. One of Leopold's first projects as a forest ranger in Arizona
and New Mexico was a campaign for the extermination of predators (chiefly wolves and mountain lions) in the interest, he then
believed, of helping the "good" animals (cattle and deer).

But the emergence of the ecological sciences brought about
a new perspective, and, as he matured, Leopold absorbed its
import. He came to believe that "the complexity of the land organism" was "the outstanding scientific discovery of the twentieth century," and he realized that predators were part of the

whole. By 1933, when he assumed a professorship of wildlife management at the University of Wisconsin, Leopold could tell his students that the entire idea of good and bad species was the product of narrow-minded human bias. One of his lecture notes stated that "when we attempt to say that an animal is 'useful,' 'ugly' or 'cruel,' we are failing to see it as part of the land. We do not make the same error of calling a carburetor 'greedy.' We see it as part of a functioning motor." On another occasion he advised those who would modify the natural world that "to keep every cog and wheel is the first precaution of intelligent tinkering." [33]

Aldo Leopold's first exploration of the ethics of the human relationship to nature appeared in a paper (unpublished until 1979) written in 1923 when he was an assistant director of the national forests in Arizona and New Mexico. Entitled "Some Fundamentals of Conservation in the Southwest," the essay began traditionally enough by assuming the need for the "development" of the region and the importance of "economic resources" in that process. For most of his paper Leopold drew on the familiar Pinchot-inspired position that conservation was necessary for continued prosperity. But in a remarkable conclusion he turned to "conservation as a moral issue." To be sure, Pinchot, Roosevelt, and especially W J McGee had used similar rhetoric, but only in speaking of equal *human* rights to resources—the familiar democratic rationale for Progressive conservation. Leopold, however, had something else in mind. The argument that the earth was man's "physical provider" and hence worthy of ethical consideration left him unsatisfied. He wondered if there was not a "closer and deeper relation" to nature based on the idea that the earth was *alive*. With this concept Leopold moved into uncharted waters. The animal rightists or humanitarians, both in England and the United States, were clearly concerned with living things. But what about geographical features such as oceans, forests, and mountains? Were they animate or inanimate, living or merely mechanical? Intuitively, Leopold rebelled against the idea of a "dead earth." He already knew enough about ecology to understand the importance of interconnections and interdependencies. Somehow this rendered hollow the traditional distinction between organic and inorganic things.[34]

Aldo Leopold's Intellectual Heritage

In his search for help with these concepts Leopold found, rather surprisingly, the Russian philosopher P. D. Ouspensky (1878–1947). Almost an exact contemporary of Leopold, Ouspensky published *Tertium Organum* in 1912. An English translation appeared in the United States in 1920, and Leopold quotes from it in his 1923 essay. The quotations are accurate, but Leopold, characteristically, provided no reference to title or page and three times misspelled the author's name as "Onspensky." Nonetheless, what excited Leopold about Ouspensky was the Russian philosopher's conviction that "there can be nothing dead or mechanical in Nature . . . life and feeling . . . must exist in everything." So, the philosopher continued, "*a mountain*, a *tree*, a *river, the fish in the river, drops of water, rain*, a *plant, fire*—each separately must possess a mind of its own." Ouspensky actually wrote about "the mind of a *mountain*," and Leopold may have remembered this phrase twenty years later when he titled his own essay, "Thinking Like a Mountain." [35]

Ouspensky based his views on the assumption that everything in the universe had a "phenomenal" or visible appearance and a "noumenal" essence. The latter was hidden to humans, and Ouspensky variously described it as life, emotions, feeling, or mind. Leopold, although a scientist, had sufficient confidence in his intuition to grasp this idea and went on to accept Ouspensky's argument that combinations of objects and processes could also be said to have lives of their own. The whole was greater than the sum of the parts. So cells functioned together to make organs, and arrangements of organs made organisms possible. But Ouspensky did not stop here. Many organisms, working together in air, water, and soil constituted a superorganism with its own particular noumenon. Such functioning communities could not be divided without destroying their collective lives, or as Ouspensky put it, in a phrase Leopold quoted, "anything indivisible is a living being." Take away the heart, for example, and you kill the greater life of the wolf. Remove the wolf from the ecosystem and you alter the noumena of the biotic community of which it was a part. The erosion of soil produces the same alteration. The conclusion Ouspensky drew and Leopold applauded was that the earth itself (or, as Leopold came to prefer, land) was not dead but alive. With his superior stylistic skills Leopold came to the Russian's aid in expressing the concept: the earth was alive, "vastly less alive than ourselves in degree, but vastly greater than ourselves in time and space—a being that was old when the morning stars sang together, and, when the last of us has been gathered unto his fathers, will still be young."[36]

For Leopold in 1923 the Ouspensky-supported assumption that the earth was, in Leopold's words, "an organism possessing a certain kind and degree of life" offered reason enough for an ethical relationship. "A moral being," he simplified the matter, "respects a living thing." This proposition, of course, could and would receive intensive scrutiny by later philosophers. But Leopold, already probing the scientific basis of a functioning earth-organism, did not pursue the philosophical puzzles. As he saw it, the "indivisibility of the earth—its soil, mountains, rivers, forests, climate, plants, and animals" was sufficient reason for respecting the earth "not only as a useful servant but as a living being."[37]

Ten years passed before Aldo Leopold wrote again about the ethical dimension of conservation. When he did so, Susan Flader thinks, he wrote "in a strikingly different manner"—as an ecologist rather than a metaphysician and theologian.[38] Her assessment is true in part. As Leopold turned in the early 1930s from a career in government to one in academe, and as he associated with renowned ecologists like Charles Elton, he absorbed a new vocabulary of chains, flows, niches, and pyramids. The glue holding the earth together consisted of food and energy circuits rather than divine forces or noumena. But there are striking continuities extending from the 1923 essay. The seeds of the key concepts in Leopold's land ethic are all present in the early paper. He had discovered the idea that a life community extended far beyond traditional definitions. He had argued for an ethical relationship to both its component parts and to the whole. And he had found that a strictly economic posture toward nature created serious ecological and ethical problems. Leopold's plunge into ecology represented not so much a switch as an extension. He never stopped working on the borderline between science and philosophy, using each to reinforce the other. When science lost sight of the broad picture in a welter of detail, philosophy adjusted the focus. Perhaps Leopold remembered Ouspensky's warning about the tendency of scientists to "always study the little finger of nature."[39]

Aldo Leopold's next step toward the land ethic was a paper read in New Mexico on May 1, 1933. Published as "The Conservation Ethic," its major contribution was the idea of ethical evolution. Like so many previous commentators on this subject, Leopold noted the parallels between human slavery and unconditional ownership of land. He hoped that the fact that slavery had been challenged and abolished would help the conservation movement. Leopold hoped that the conservation movement represented an awareness that "the destruction of land . . . is wrong." And, Leopold made clear, he did not mean "wrong" in the sense of inexpedient or economically disadvantageous. He meant it in the same sense that abuse of another human being was wrong.[40]

Nowhere in the 1933 essay did Leopold refer to the *rights of* land or nonhuman life although Muir, Moore, and Evans had all advanced that concept. Ethics for Leopold were the ideas or ideals of people that acted as restraints on people. He defined an ethic as a "limitation on freedom of action in the struggle for existence." In other words, ethics applied to situations where a person who could have done a particular action held back because he knew that action was wrong. Sometimes, Leopold understood, this involved working directly against immediate self-interest. The ethical person forgoes the opportunity to improve his economic position by robbing another person. In the same way, Leopold hoped that a land ethic might be a constraint against robbing or exploiting the land. Although subsequent philosophers have paid extensive attention to the point, Leopold simply dismissed the notion that animals, plants, and soil had reciprocal ethical obligations toward people. For Leopold it was a one-way street: human beings were the ones to exercise restraint, to extend *their* ethics to include nature.[41]

Aldo Leopold gave no specific indication in his 1933 statement (or, for that matter, in *A Sand County Almanac*) that anyone had ever thought about expanding ethical sequences before his time.

Yet he must have known that Darwin had, in 1871, written extensively on the subject. In fact, in identifying "the tendency of interdependent individuals or societies to evolve modes of cooperation" known as ethics, Leopold nearly plagiarized the great English biologist. He also ignored the anticipation of his ideas by William Lecky, John Muir, Edward Evans, J. Howard Moore, Henry Salt, Liberty Hyde Bailey, and Albert Schweitzer. Yet the work of these men was readily available when Leopold was in college or early in his professional career. Granted that as an ecologist Leopold took his ethics further than most of these thinkers—to collections of organisms and habitats organized as ecosystems—but it is disconcerting that this scientist, so meticulous in his recording of biological facts, would play so loosely with historical ones. Similarly surprising is the occasional tendency of Leopold scholars to aggrandize their subject at the expense of historical accuracy. J. Baird Callicott, for example, overstates the case with his opinion that Leopold's ideas are "the first self-conscious, sustained, and systematic attempt in modern Western literature to develop an ethical theory which would include non-human natural entities and nature itself in the purview of morals."[42]

In late 1947 and early 1948 Leopold reviewed his 1923 and 1933 essays, added insights from subsequent papers of 1939 and 1947, and wrote a final chapter for *A Sand County Almanac* entitled "The Land Ethic." Ethics, he explains, derive from the recognition that "the individual is a member of a community of interdependent parts" (203–4). On the one hand the individual competes within this community, but "his ethics prompt him also to co-operate (perhaps in order that there may be a place to compete for)" (204–5). The land ethic, then, "changes the role of *Homo sapiens* from conqueror of the land-community to plain member and citizen of it. It implies respect for his fellow-members, and also respect for the community as such" (204). Behind this sentence lies Leopold's recognition that while, in one sense, humans are simply members of a "biotic team," in another their technologically magnified power to affect nature sets them apart from the other members (205). So human civilization needed the restraints afforded by a land ethic. Just as concepts of right and

Aldo Leopold's Intellectual Heritage

wrong had made human society more just, Leopold felt they would enhance justice among species and between man and the earth. The entire import of *A Sand County Almanac*, Leopold writes in his foreword, is directed to helping land "survive the impact of mechanized man" (viii). His statement of the basic problem and solution is characteristically pithy and powerful: "We abuse land because we regard it as a commodity belonging to us. When we see land as a community to which we belong, we may begin to use it with love and respect" (viii).[43]

Frequently in *A Sand County Almanac*, as well as in his other writings, Leopold takes an instrumental view of the land ethic. It embodies restraints that help humans live a healthy life. It is prudent to be ethical with regard to the natural order that sustains the human one. A battle-scarred veteran of conservation policy wars, Leopold knew this was the best way to sell his philosophy in the 1930s and 1940s. He also knew it was not the full story. His most radical ideas, and his greatest significance for the 1960s and beyond, concern the *intrinsic* rights to existence of nonhuman life forms and of life communities or ecosystems. Early in "The Land Ethic" Leopold affirms the "right to continued existence" not only of animals and plants but of waters and soils as well (204). And he writes that the life forms that share the planet with people should be allowed to live "as a matter of biotic right, regardless of the presence or absence of economic advantage to us" (211). This means "there are obligations to land over and above those dictated by self-interest," obligations grounded on the recognition that humans and the other components of nature are ecological equals (209).

This was the intellectual dynamite in *A Sand County Almanac*. The Darwinian evolutionists and the old-style humanitarians had occasionally glimpsed the idea of a morality that extended beyond human society, but Leopold, with the aid of ecology, gave it its most dramatic articulation at least up to the late 1940s. Most of the earlier advocates of extended ethics (with the exception of J. Howard Moore, who also included "life processes") dealt almost exclusively with individual organisms and then, generally only with the higher animals. This is not to denigrate them. Theirs was an expected intellectual way-station on the road from anthropocentric to ecocentric ethics. Leopold's achievement was to follow the road to its logical termination in ecosystems or the environment. J. Baird Callicott, as it turns out, rightly characterizes him as the father of *environmental* ethics.

If Darwin killed dualism, the ecologists presided over its burial. Humans were simply one of many members of a greatly expanded biotic community. The moral implications of this idea for human behavior were, to say the least, problematic, and philosophers after Leopold would devote hundreds of pages to the subject. But Leopold was quite clear about what he thought the land ethic mandated in terms of behavior. It did not, at the outset, mean having *no* impact on one's environment. As a biologist Leopold knew this was an impossibility for any organism, and he might have chuckled at vegetarians such as Henry Salt and extreme right-to-life sects such as the Jains who, following the philosophy of *Ahimsa,* breathed through gauze so as not to inhale living microorganisms. Even Albert Schweitzer's aid to struggling worms and insects would have struck Leopold as naive and beside the main point of land health. Leopold's concept of reverence-for-life was precisely that—for life in toto and not so much for the individual players in the process. Leopold would have approved of Schweitzer's principle of taking life only for essential purposes and, then, with reverence for that which was killed. Hunting, meat eating, even, in Leopold's words, "the alteration, management and use" of the ecosystem were necessary and inevitable. The essential proviso, Leopold wrote as early as 1933, was that any human action be undertaken in such a way as "to prevent the deterioration of the environment." [44] By 1948, when he finished "The Land Ethic" for *A Sand County Almanac,* Leopold had refined this principle into what has become his most widely quoted precept. A land-use decision "is right when it tends to preserve the integrity, stability, and beauty of the biotic community. It is wrong when it tends otherwise" (224–25).

Leopold was well aware of the massive obstacles standing in the way of ethically directed human relations to the environment. "No important change in ethics," he wrote in *A Sand County Almanac,* "was ever accomplished without an internal change in our intellectual emphasis, loyalties, affections, and convictions."

The conservation movement of his day had not, in his view, touched these "foundations of conduct." As proof, he submitted that "philosophy and religion have not yet heard" of "the extension of social conscience from people to land" (209–10). In this belief Leopold was both correct and in error. Philosophy, biology, history, religion, and even law (the humane legislation) had all heard, at least, of the extension of ethics. But Leopold was correct in assuming that Western thought in general contained little approaching the holistic character of his moral philosophy. Still, in matters of slowly shifting attitudes and values, he knew the value of patience. Ethics, after all, were ideals, not descriptions of how people actually behaved: "We shall never achieve justice or liberty for people. In these higher aspirations the important thing is not to achieve, but to strive." [45]

The parallel Leopold drew between human-to-human and human-to-nature ethics informed many of his final essays. In "The Ecological Conscience" of 1947 he noted that it "has required 19 centuries to define decent man-to-man conduct and the process is only half done; it may take as long to evolve a code of decency for man-to-land conduct." His prescription for such progress was not to allow economics to dictate ethics: "Cease being intimidated by the argument that a right action is impossible because it does not yield maximum profits, or that a wrong action is to be condoned because it pays." "That philosophy," Leopold concluded, "is dead in human relations, and its funeral in land-relations is overdue." [46]

Aldo Leopold's pessimism concerning public comprehension, not to speak of acceptance, of the land ethic was supported by the early history of *A Sand County Almanac*. As an unpublished typescript it was sent to and rejected by so many publishing houses that the author despaired of ever seeing his work in print. Leopold did not live to see the reviews of *Sand County*, but they would probably have disappointed him. Most critics understood the book to be just another collection of charming nature essays. Very few reviewers even recognized the thesis that a later generation would find so compelling. Initial sales of the slender green volume were slow, and, before its renaissance in the 1960s, it sold scarcely twenty thousand copies.

The most obvious reason for the initial lack of public interest in Leopold's ideas was their truly radical quality. His proposals called for a complete restructuring of basic American priorities and behavior, and a radical redefinition of progress. The conquest and exploitation of the environment that had powered America's westward march for three centuries was to be replaced, as an ideal, by cooperation and coexistence. The land ethic placed unprecedented restraints on a process that had won the West and lifted the nation to at least temporary greatness as a world power. Taken literally, Leopold's philosophy would abruptly put an end to the accustomed freedom with which Americans had hitherto dealt with nature. It was no longer to be a master-slave relationship; the land had rights too.

Notes

1. Donald Worster's *Nature's Economy: The Roots of Ecology* (San Francisco: Sierra Club Books, 1977) is the definitive history of ecology as both science and philosophy. Older, but still useful, summaries are W. C. Allee et al., *Principles of Animal Ecology* (Philadelphia: Saunders, 1949), 1–72; Richard C. Brewer, "A Brief History of Ecology," *Occasional Papers of the C. C. Adams Center for Ecological Studies* 1 (1960); and Charles S. Elton, *The Pattern of Animal Communities* (London: Methuen, 1966), 29–44. Joseph V. Siry, *Marshes of the Ocean Shore: Development of an Ecological Ethic* (College Station, Tex.: Texas A&M Press, 1984) is a useful review of the contributions of ecologists to the understanding of the natural world.

2. Commentary on Transcendentalism may be found in Sherman Paul, *The Shores of America: Thoreau's Inward Exploration* (Urbana: University of Illinois Press, 1958); Worster, *Nature's Economy;* and Roderick Nash, *Wilderness and the American Mind* (New Haven: Yale University Press, 1967; third edition, 1982), ch. 5.

3. *The Writings of Henry David Thoreau,* ed. Bradford Torrey, vol. 9 (Boston: Houghton Mifflin, 1906), 210.

4. Henry David Thoreau, *The Maine Woods,* ed. Joseph J. Moldenhauer (Princeton: Princeton University Press, 1972), 181.

5. *Writings of Thoreau,* vol. 11, 450.

6. Thoreau, *Maine Woods,* 121.

7. *Writings of Thoreau,* vol. 4, 422; vol. 3, 381.

Aldo Leopold's Intellectual Heritage

8. *Writings of Thoreau*, vol. 10, 51.

9. George Perkins Marsh, *Man and Nature: Or, Physical Geography as Modified by Human Action* (Cambridge: Harvard University Press, 1965), 36, 45, 46, 91–92. David Lowenthal, *George Perkins Marsh: Versatile Vermonter* (New York: Columbia University Press, 1958) is the definitive biography.

10. *John of the Mountains: The Unpublished Journals of John Muir*, ed. Linnie Marsh Wolfe (Boston: Houghton Mifflin, 1938), 138. Muir's life and thought are interpreted in Stephen Fox, *John Muir and His Legacy: The American Conservation Movement* (Boston: Little, Brown, 1981), 43–44; Edith Jane Hadley, *John Muir's Views of Nature and Their Consequences* (Ph.D. diss., University of Wisconsin, 1956); Nash, *Wilderness*, ch. 8; and Michael Cohen, *The Pathless Way: John Muir and the American Wilderness* (Madison: University of Wisconsin Press, 1984).

11. John Muir, *Our National Parks* (Boston: Houghton Mifflin, 1901), 57–58.

12. John Muir, *A Thousand-Mile Walk to the Gulf*, ed. William F. Badè (Boston: Houghton Mifflin, 1917), 324, 356.

13. John Muir, *My First Summer in the Sierra* (Boston: Houghton Mifflin, 1911), 211. The book was a composite of Muir's much earlier journals, somewhat reworked. The first draft of this statement, written July 27, 1869, just after Muir had his first view from the rim of Yosemite Valley, seems even more striking: "When we try to pick out anything by itself, we find that it is bound fast by a thousand invisible cords that cannot be broken to everything in the universe. I fancy I can hear a heart beating in every crystal, in every grain of sand and see a wise plan in the making and shaping and placing of every one of them. All seems to be dancing in time to divine music." (As quoted in Fox, *John Muir*, 291.)

14. *The Life and Letters of Charles Darwin*, 2 vols., ed. Francis Darwin (New York: Appleton, 1888), vol. 1, 368.

15. Charles Darwin, *The Descent of Man and Selection in Relation to Sex* (New York: Appleton, 1874), 81, 138, 140. Useful secondary literature includes Worster, *Nature's Economy*, pt. 3; Gavin de Beer, *Charles Darwin* (Garden City, N.Y.: Doubleday, 1964); Gertrude Himmelfarb, *Darwin and the Darwinian Revolution* (Gloucester, Mass.: P. Smith, 1967).

16. William E. H. Lecky, *History of European Morals from Augustus to Charlemagne*, 2 vols., (London: Longmans, Green, 1869), vol. 1, 143, 103; Darwin, *Descent of Man*, 140.

17. J. Baird Callicott, "Animal Liberation: A Triangular Affair," *Environmental Ethics* 2 (Winter 1980): 311–38. The quote, by Kenneth Goodpaster, appears on page 315. Callicott has softened his position on the

incompatibility of animal liberation and environmental ethics in a book review of Mary Midgeley's *Animals and Why They Matter* (Athens, Ga.: University of Georgia Press, 1983) in *Canadian Philosophical Review* 10 (1986): 464–67. A balanced assessment of the relationship between the humane movement and environmental ethics, critical of Callicott, is Mary Anne Warren, "The Rights of the Nonhuman World," in *Environmental Philosophy: A Collection of Readings*, ed. Robert Elliot and Arran Gare (University Park: Pennsylvania State University, 1983), 109–34. See also Marti Kheel, "The Liberation of Nature: A Circular Affair," *Environmental Ethics* 7 (1985): 135–49.

18. Mark Sagoff, "Animal Liberation and Environmental Ethics: Bad Marriage, Quick Divorce," *Osgood Hall Law Journal* 22 (1984): 297–307.

19. Evans' *Popular Science Monthly* essay appeared in his book *Evolutional Ethics and Animal Psychology* (New York: Appleton, 1897), 82–104; the references here are 83, 88–99, and 91. The essay may also be found in an abridged form in Donald Worster, ed., *American Environmentalism: The Formative Period, 1860–1915* (New York: Wiley, 1973), 198–208. White's essay is "The Historical Roots of Our Ecological Crisis," *Science* 155 (1967): 1203–7.

20. Evans, *Evolutional Ethics*, 99–100.

21. J. Howard Moore, *The Universal Kinship* (London: Bell and Sons, 1906), 107, 239, 297; Moore, *The New Ethics*, rev. ed (Chicago: S. A. Block, 1909), 215.

22. Moore, *Universal Kinship*, vii, 324; Moore, *New Ethics*, 15, 19.

23. Moore, *Universal Kinship*, 324, 273; Moore, *New Ethics*, 169. The emphasis is in the original work.

24. Moore, *Universal Kinship*, 281, 324.

25. Ibid., 329, viii; Moore, *New Ethics*, 13. Emphasis in original.

26. Evans, *Evolutional Ethics*, 164.

27. Liberty Hyde Bailey, *The Holy Earth* (New York: Charles Scribner's Sons, 1915), 14, 24, 30–31. For secondary analysis see Philip Dorf, *Liberty Hyde Bailey: An Informal Biography* (Ithaca: Cornell University Press, 1956); Stephen Fox, "Liberty Hyde Bailey: The Earth as Whole, The Earth as Holy," *Orion* 2 (Autumn 1983): 13–23; and Andrew Denny Rodgers: *Liberty Hyde Bailey: A Story of American Plant Science* (Princeton: Princeton University Press, 1949), 405ff.

28. Albert Schweitzer, *Out of My Life and Thought: An Autobiography* (New York: Holt, 1933), 185, 186. The most recent biography is James Brabazon, *Albert Schweitzer* (New York: Putnam, 1975). See also George Seaver, *Albert Schweitzer: The Man and His Mind* (London: A. and C. Black, 1969); and Norman Cousins, *Doctor Schweitzer of Lambarene* (New York: Harper and Row, 1960).

Aldo Leopold's Intellectual Heritage

29. Schweitzer, *Out of My Life*, 188.

30. Albert Schweitzer, *Philosophy of Civilization: Civilization and Ethics*, trans. John Naish (London: A. and C. Black, 1923), 254.

31. Albert Schweitzer, *Indian Thought and Its Development*, trans. C. E. B. Russell (New York, 1936), 261–62; Albert Schweitzer, *The Animal World of Albert Schweitzer*, ed. Charles R. Joy (Boston: Beacon, 1950), 169.

32. Stewart L. Udall, *The Quiet Crisis* (New York: Holt, Rinehart & Winston, 1963), 206; Callicott, "Animal Liberation: A Triangular Affair," 311; Wallace Stegner, "Living on Our Principal," *Wilderness* 48 (Spring 1985): 15; Donald Fleming, "Roots of the New Conservation Movement," *Perspectives in American History* 6 (1972): 18; Clay Schoenfeld, "Aldo Leopold Remembered," *Audubon* 80 (May 1978): 79; Van Rensselaer Potter, *Bioethics: Bridge to the Future* (Englewood Cliffs, N.J.: Prentice-Hall, 1971), v.

33. Aldo Leopold, "Wherefore Wildlife Ecology?" undated lecture notes, Aldo Leopold Papers, University of Wisconsin Archives, Madison, Wis., box 8; Aldo Leopold, *Round River*, ed. Luna Leopold (New York: Oxford University Press, 1953), 146–47. Biographical attention may be found in Roderick Nash, *Wilderness and the American Mind*, ch. 11; Roderick Nash, "Aldo Leopold," *Dictionary of American Biography*, vol. 4 (New York: Scribner, 1974), 482–84; Fleming, "Roots of the New Conservation," 18–27; Worster, *Nature's Economy*, 271–74, 284–90; and especially in Susan L. Flader, *Thinking Like a Mountain: Aldo Leopold and the Evolution of an Ecological Attitude toward Deer, Wolves, and Forests* (Columbia: University of Missouri Press, 1974). Flader's *The Sand Country of Aldo Leopold* (San Francisco: Sierra Club Books, 1973) also contains biographical data as does Boyd Gibbons, "Aldo Leopold: A Durable Scale of Values," *National Geographic* 160 (November 1981): 682–708. Curt Meine's forthcoming book to be published by the University of Wisconsin Press, will be the definitive biography. See also Curt Meine, "Aldo Leopold's Early Years" in this volume.

34. Leopold's 1923 essay was finally published, thanks to the efforts of Eugene C. Hargrove, editor of *Environmental Ethics*, as "Some Fundamentals of Conservation in the Southwest" in *Environmental Ethics* 1 (1979): 131–41.

35. P. D. Ouspensky, *Tertium Organum: The Third Canon of Thought, a Key to the Enigmas of the World*, trans. E. Kadloubovsky (New York: Knopf, 1981), 166. Emphasis in original.

36. Ibid., 118, 166, 168; Leopold, "Some Fundamentals," 140.

37. Leopold, "Some Fundamentals" 139–40.

38. Susan L. Flader, "Leopold's *Some Fundamentals of Conservation:* A Commentary," *Environmental Ethics* 1 (1979): 143.

39. Ouspensky, *Tertium Organum,* 179.

40. Aldo Leopold, "The Conservation Ethic," *Journal of Forestry* 31 (1933): 634, 635, 640.

41. Leopold, "Conservation Ethic," 634. Among the more perceptive recent dissections of Leopold's philosophy on these points are J. Baird Callicott, "Elements of an Environmental Ethic: Moral Considerability and the Biotic Community," *Environmental Ethics* 1 (1979): 71–81; Scott Lehmann, "Do Wildernesses Have Rights?" *Environmental Ethics* 3 (1981): 129–46 and Worster, *Nature's Economy,* 284ff.

42. Leopold, "Conservation Ethic," 634; J. Baird Callicott, "The Land Aesthetic," *Environmental Review* 7 (1983): 345.

43. The later essays that informed "The Land Ethic" chapter were "A Biotic View of Land," *Journal of Forestry* 37 (1939): 727–30 and "The Ecological Conscience," *Bulletin of the Garden Club of America* (September 1947), 45–53. See Curt Meine, "Building 'The Land Ethic,'" in this volume for a discussion of the preparation of Leopold's most important statement.

44. Leopold, "Conservation Ethic," 641. Leopold's first published version of this precept appeared in "The Ecological Conscience," 52, and it stressed the common membership shared by humans and other life forms in one community. The 1947 version reads: "A thing is right only when it tends to preserve the integrity, stability, and beauty of the community, and the community includes the soil, waters, fauna, and flora, as well as people." Extensive analysis of this idea appears in James D. Heffernan, "The Land Ethic: A Critical Appraisal," *Environmental Ethics* 4 (1982): 235–47 and in Tom Regan, ed., *Earthbound: New Introductory Essays in Environmental Ethics* (Philadelphia: Temple University Press, 1984), 268ff., 351ff. Regan, who disagrees with Leopold's deemphasis on the rights of individual organisms, comments briefly on his thought in *The Case for Animal Rights* (Berkeley: University of California Press, 1983), 361ff. See J. Baird Callicott's review of Tom Regan's *The Case for Animal Rights* in *Environmental Ethics* 7 (1985): 365–72.

45. Leopold, *Round River,* 155.

46. Leopold, "The Ecological Conscience," 53. Underscoring the importance of ecology to his ethical system, Leopold used the term "ecological conscience" here to describe what in *A Sand County Almanac* he calls "the land ethic."

II * The Book

4 ⋆ The Making of *A Sand County Almanac*

DENNIS RIBBENS

The evolution of the text of *A Sand County Almanac* and of Aldo Leopold's thinking about the substance and the structure of the text during the time of its formation (from November 1941 to April 1948) is discernible in the correspondence, essay drafts, and other manuscripts contained in the Leopold Papers at the University of Wisconsin–Madison Archives. Especially in the letters between Leopold and his publishers and his friends one can trace the formation of the concept of the book. In them one finds the debate over what constitutes a nature book, the debate about the right interplay between nature observation and ecological preachment. In them one observes Leopold working out his own answers to these questions. Here I examine the period during which *A Sand County Almanac* evolved both as concept and structure in order to discover the development of Leopold's thinking about what a nature book should be. I will also briefly comment on some of the mechanical aspects of writing, revising, and editing the work: changes within the essays, changes in the type of essays, changes in the titles, and changes in organization.

This essay was originally published in the *Wisconsin Academy Review*, volume 28, number 4 (September, 1982): 8–10. It has been revised for publication in this volume. Sources of the archival materials and full citations for published works mentioned here are provided in a note on the sources at the end of the essay.

Prior to 1941 Leopold had published essays in many journals, some as early as the 1910s. A portion of the most frequently quoted essay in *A Sand County Almanac*, "The Land Ethic," for example, first appeared in 1933 as "The Conservation Ethic." "Conservation Esthetic" first appeared in 1938. Most of Leopold's early essays were either technical or, like the two just cited, overtly exhortative—"philosophic essays" Leopold called them. It is important to keep in mind the kinds of essays Leopold wrote before 1941 if one is to understand the controversy within Leopold himself and between Leopold and his correspondents during the 1941–1947 period in which the book was shaped.

A few of the 1930s essays, like "Marshland Elegy" (1937), shared with the philosophic essays their conservation and ecological purposes, but were presented in mostly descriptive and narrative terms. None of these essays, however, was intended to be merely or primarily narration or description. In them, event was in part substituted for reason as ecological argument. Several essays which appeared in 1941, and which were later incorporated into *A Sand County Almanac,* anticipate the kinds of essays Leopold was later to write.

The only pre–November 1941 unpublished essay to use the shack experience was "65287," later changed to "65290" (why I cannot say). "Bur Oak is Badge Of Wisconsin" (an early and shorter version of "Bur Oak"), "The Geese Return," "Sky Dance," and "The Plover is Back From Argentine" all appeared in *Wisconsin Agriculturalist and Farmer* during 1941 and 1942 and predate Leopold's thinking about *A Sand County Almanac.* Although these early versions anticipate the tone of the later shack essays (indeed some of them were revised in that direction several years later), these essays were initially thought of as conservation essays for farmers.

A letter to Leopold from Alfred A. Knopf publishers dated November 26, 1941, stimulated Leopold's serious consideration of the kind of nature essays he should write and of how they might be coherently organized. With that letter the making of *A Sand County Almanac* began. In it Harold Strauss told Leopold that Knopf sought someone to write "a personal book recounting adventures in the field . . . warmly, evocatively, and vividly written. . . . a book for the layman . . . [with] room for the author's opinions on ecology and conservation . . . worked into a framework of actual field experience."

Notice that at the very outset of the Knopf-Leopold correspondence the fundamental variables which make up nature writing not only were identified but also were couched within an inevitable tension. On the one hand, Strauss says that a nature book must be personal and narrative—a recounting of field adventures and experiences, informed, warm, evocative. On the other hand, a nature book might contain the author's opinions, his considered analysis of nature, his comment about natural events and man's place in them. It might address ecological and conservationist matters. And notice too his insistence that such ecological considerations be "worked into a framework of actual field experience," not the reverse. That was not the approach Leopold's essays had been taking up to that time. Conservation issues, not descriptions of nature, were the controlling element in his essays.

Ironically, had Knopf's editors held to their desire to mold ecological considerations into field experience, they would have judged Leopold's essays more favorably in 1944 and in 1947. Leopold's response (December 3, 1941) addressed this issue, an issue which was to haunt him for the next six years. In it Leopold also questioned "how far into ecology (that is, how far beyond *mere natural history*) such a book should attempt to go." "I am convinced," he wrote, "that the book should go part way into ecological observation" (emphasis added).

This matter of what constitutes a nature book, the right mix of nature observation and conservation exposition, received little attention from Leopold during the next two or more years. On December 29, 1941, Leopold wrote to Knopf, "I am out as sole author for a year or two." He claimed to be writing "a series of ecological essays . . . as a Christmas book." I have uncovered no other reference to this unusual Christmas-book idea. Most probably Leopold had in mind a collection of the essays which he was then writing for the *Wisconsin Agriculturalist and Farmer*.

It is worth noting that Leopold described what he was writing in 1941 as "ecological essays," a term appropriate for the largely nonnarrative, exhortative conservation pieces written before that time. In January of 1942 Knopf asked to see some of these ecological essays. Leopold pleaded that upon closer consideration he found the essays not to be ready. In April of 1943 Knopf again asked how the essays were progressing. Leopold responded that he would get to them in the next year or two.

In January of 1944 Knopf once again inquired about the essays. This time Leopold was able to say, "I have been working steadily" (January 28, 1944). And he had been busy, both writing and revising. Of the datable manuscript drafts of the essays from *A Sand County Almanac* found in the Leopold Papers of the University of Wisconsin–Madison Archives, more than two dozen date from 1943 and the first half of 1944, the period from September of 1943 through June of 1944 being the most productive. Probably more than a dozen of the book's forty-one essays were written during that period.

That period was rich in correspondence between Leopold and his friend and former student, Hans Albert Hochbaum, whom Leopold had helped to organize a wild-fowl research station at Delta, Manitoba. Hochbaum was to do the illustrations and Leopold the essays for the book they were planning for Knopf. Their letters to one another reveal what essays Leopold was then working on, what tone and content he sought for them, and especially what overall effect Leopold wanted the collection to have. As early as May 7, 1943, in a letter to Hochbaum, Leopold spoke of "our joint venture." "Let's by all means reinstate the original plan," he wrote, "and keep sending each other whatever materials we manage to bring together." Regarding the book and its drawings, Leopold wrote, "[t]his is a personal venture, and I take special pride in its 'home-made' aspect" (June 18, 1944). Of Hochbaum's critical advice Leopold said, "I am learning a lot from your letters" (June 3, 1944). Later he wrote about their association, "our intellectual partnership is one of the anchors of my ship. Without it I would be adrift" (October 17, 1947).

In his letters to Hochbaum, Leopold expressed concern over

what he called the "literary effect" of the essays. Regarding his effort to reconcile the need, on the one hand, to provide enough environmental data to permit ethical judgment and, on the other, to achieve a satisfactory artistic or literary effect, he wrote at length in a letter to Hochbaum dated March 1, 1944.

> When you paint a picture, it conveys a single idea, and not all of the ideas pertinent to the particular landscape or action. If you inserted all of the ideas of your picture, it would spoil it.
>
> In order to arrive at an ethical judgment, however, about any question raised by the picture, you need to consider all pertinent ideas, including those which changed in time. It seems to me, therefore, that any artistic effort, whether a picture or an essay, most often contains less than is needed for an ethical judgment. That is approximately what I meant when I said I intended to revise the essays insofar as could be done without spoiling the literary effect. . . . I do know that the essays can give a more accurate judgment, particularly in reference to my own changes of attitude in time without hurting literary effect, and possibly improving the literary effect.

These letters of 1943 and early 1944 refer to "The Green Lagoons," "Too Early," "Illinois Bus Ride," "Draba," "Marshland Elegy," "Escudilla," "sketch of the chickadees," "The Flambeau," "Odyssey," "Great Possessions," "Thinking Like a Mountain," and "Pines Above the Snow." Hochbaum, in a letter Leopold marked "important letter," combined praise and criticism of the essays, and encouraged Leopold to worry less about "literary effects." "Since you can give a lilt to the deadest subject," he wrote, "it seems to me that [the quality of the essay] is in what you are writing about, not in your technique" (March 11, 1944).

Hochbaum was able to identify precisely those issues of unity, tone, and emphasis that were to plague Leopold and his prospective publishers for years. He found the overarching theme of the essays hard to uncover. He considered Leopold's tone elitist and cynical. He encouraged Leopold to write more simply, personally, optimistically. He further suggested that Leopold, in his struggle to get the right mix of natural facts and "ethical judgment" (March 11, 1944) think of the series of essays as a self-portrait, and that the Leopold depicted be "less a person than he is a Standard" (March 11, 1944), but a standard which finds lessons in his own life as well as in the lives of others. As Hochbaum wrote on February 4, 1944:

> The lesson you wish to put across is the lesson that must be taught—preservation of the natural. Yet it is not easily taught if you put yourself above other men. That is why I mentioned your earlier attitude toward the wolf. The Bureau Chief had as much right to believe we should be rid of the Escudilla bear, or the government crews to plan roads for the crane marsh, as you had the right to plan the extermination of wolves in New Mexico. One gathers from parts of "Escudilla" and "Marshland Elegy" that you bear a grudge against these fellows for not thinking as you when, in your own writings, you show that you once followed a similar pattern of thought. Your lesson is much stronger, then, if you try to show how your own attitude towards your environment has changed.

Leopold himself had in an earlier letter to Hochbaum acknowledged that "about the question of attitude in the essays—we all go through the wringer at one time or another" (January 29, 1944). A month later on the same matter, Hochbaum said, "You have sometimes followed trails like anyone else that lead you up wrong alleys. That is why I suggested the wolf business. . . . I hope you will have at least one piece on wolves alone" (March 11, 1944). On March 21, Leopold said he planned to write a wolf essay soon. On April 4, he wrote, "I am roughing out an essay or two, working toward your idea of a shack series." Leopold also enclosed a draft of "Thinking Like a Mountain," an essay of confession to counter the stridency of his 1930s ecological essays. Said Hochbaum, "'Thinking Like a Mountain' fills the bill perfectly" (April 15, 1944).

On June 6, 1944, Leopold sent thirteen essays both to the Macmillan Company, who had by then also contacted Leopold, and to Knopf. Much can be learned from a close examination of the list of those essays and of Leopold's comments about it.

1. Marshland Elegy
2. Song of the Gavilan
3. Guacamaja
4. Escudilla

In his cover letter Leopold said, "The object, which should need no elaboration if the essays are any good, is to convey an ecological view of land and conservation."

Knopf had initially asked for "a personal book recounting adventures in the field" (November 26, 1941). A first glance at this list might lead one to conclude that Leopold had attempted no accommodation between philosophical ecological essays and mere natural history. Upon closer inspection, however, one can see that he did. Conspicuously absent from the list are three non-narrative ecological essays written between 1932 and 1943: "Conservation Esthetic," "The Conservation Ethic," and "Wildlife in American Culture." Clearly Leopold's purpose in these thirteen essays (two more—"The Flambeau" and "Clandeboye"—were added in August of 1944), upon which no organizational structure had yet been imposed, was to popularize and to dramatize through narrated events, in some cases based upon his own personal experiences, those same ecological and conservation issues which he had addressed in the earlier philosophic essays.

But if, on the one hand, the more philosophical essays do not appear, neither, on the other, do the essays based on his shack experiences, except for incidental mention. It is interesting to note that after ten years at the shack and after two and a half years of serious thought about nature essays, Leopold mentioned the shack experience in only two of the fifteen essays, "Great Possessions" and "Pines Above the Snow." Unlike the 1948 draft, the 1944 draft of "The Geese Return" contained no shack reference. Ecological preachment, made accessible to the public by means of described events and experiences, is the dominant essay approach by 1944. "Draba" is the most notable exception, a gentle, elegant description carrying only an implicit ecological argument.

This change in the perspective of Leopold's essays is evident in what he said in his cover letter about a title for the book. "I once thought to call it 'Marshland Elegy—And Other Essays,' but 'Thinking Like a Mountain—And Other Essays,' now strikes me as better" (June 6, 1944). "Marshland Elegy," which dates from 1937, begins with an exquisitely poetic portrayal of a marsh dawn, and ends with a harsh attack on governmental blundering and the prospect of an ecological doomsday. In that essay, Leopold stands as aloof critic, the judge of what is right. By contrast "Thinking Like a Mountain" is personal, experiential, humble, even confessional. It records Leopold's own ecological blunders. More profound than "Elegy," it quietly speaks of individual attitude, of Leopold's own change in attitude. In place of the doomsday ending in the former, the latter concludes with his own emendation of Thoreau's hopeful dictum, "In wildness is the salvation [Thoreau had written "preservation"] of the world."

This combining in the same essay of wolf description, personal experience, attitude change, and ecological comment troubled the editors at Knopf. Their letter of rejection of July 24, 1944 (Macmillan had rejected the essays with virtually no comment a few days earlier) triggered Leopold's struggle, that would continue for the next three and a half years, to define to his own satisfaction what a nature book should be. The comments included in Knopf's letter of rejection and Leopold's response to them are critical to an understanding of the concept and structure of *A Sand County Almanac,* a monumental classic of modern nature writing. For that reason Knopf's entire July 24, 1944, letter follows.

Dear Professor Leopold,

We have discussed your essays here and find that, while we like your writing, they do not seem altogether suitable for book publication in their present form. One reason is that they are so scattered in subject matter, and it also seems to us that the point of view and even the style varies from one essay to another. Pieces of only a page or two in length are also rather difficult to put into a book. And of course the dozen articles submitted would make a very slim volume indeed. I am sure you plan these only as part of a volume.

I wonder if you would consider making a book purely of nature observations, with less emphasis on the ecological ideas which you have incorporated into your present manuscript? It seems to us that these ecological theories are very difficult indeed to present successfully for the layman. Certainly, the repetition in chapter after chapter of a book, of the idea that the various elements and forces of nature should be kept in balance would end by becoming monotonous. Would it not be better to make the greater part of the book observation of wild life in narrative form, such as your pieces on "Great Possessions" and "The Green Lagoons," adding a chapter developing the ecological interpretations?

Such a book should, we feel, be based on your own experiences and if possible should be limited to one region of the country. In the present collection, we feel a distinct break between the middle-western and southwestern essays, because of the completely different conditions existing in the two regions. Some sort of unifying theme or principles must be found for a book of this sort, we think, and perhaps it would hold together better if it were limited to a single part of the country.

One reason the ideas about the balance of nature, as embodied in these essays, do not seem successfully presented is, I feel, that the reader is apt to get a confused picture of what you advocate. Sometimes it seems that you want more intelligent planning, but you point out that nature's balance was upset with the coming of civilization, and you certainly do not seem to like the ordinary brand of conservationists and government planners. I think the average reader would be left somewhat uncertain as to what you propose. Perhaps in a single essay, all these ideas could be related so that your basic theme would become clearer.

I should add that we are impressed with your writing, with the freshness of observation which it reflects, and the skill of phrase. We believe that readers who like nature will enjoy such writing and hope that we can work out with you a successful plan for a volume. I would appreciate hearing your reaction to the above, and will hold the essays until you tell me what to do with them.

Yours Sincerely,
Clinton Simpson

That the essays were yet not organized into a book form is true. But the more basic issue remained the direction Leopold's essays took, their effort to combine narrative and exposition. Not surprisingly Knopf liked "The Green Lagoons" and "Great Possessions." The editors still preferred "a book purely of nature observations, with less emphasis on ecological ideas." The heart of the Knopf-Leopold debate was the perceived conflict between observation of nature and comment about nature, between aesthetic response and ethical insight, between nature as other and man's interaction with it. Knopf wanted the "what-I-saw-while-in-the-woods" sort of nature book. Leopold's concerns by contrast were planetary and ethical as well as provincial and descriptive.

In a letter dated August 24, 1944, Knopf, after seeking the judgment of two unnamed professional writers, dropped its concern for regional focus, but persisted in demanding more essays, longer essays, elimination of repetitive ecological arguments, and the addition of a chapter "which sums up the argument for the forces of nature." Leopold's reply to Knopf's letter of rejection made it clear that his agenda was "conservation in continental rather than in local terms" (July 27, 1944). But he agreed with Knopf that the essays should, whenever possible, be presented in narrative form.

Hochbaum's reply to the Knopf letter of July 24, which Leopold at once shared with him, pointed out the similarities between Knopf's and Hochbaum's previous criticisms. He urged Leopold to recast the entire book around the shack experience—something narrative, closer to nature, more hopeful in tone (July 31, 1944). Earlier, Leopold had written Hochbaum saying, "I am roughing out an essay or two, working toward your idea of a shack series" (April 4, 1944). On August 17, 1944, Leopold admitted to Hochbaum that "the shack essays . . . are of a different cast than the others." By the end of the summer Leopold pledged to redo the essays along the lines Knopf suggested—longer essays with a discernible difference between the body of natural observation and the final section on ecological matters. A few months later he wrote to Hochbaum, "I think I catch the idea of the new book and it sounds very good to me.

The Making of *A Sand County Almanac*

I have been flirting with the almanac idea myself as a means of giving 'unity' to my scattered essays" (November 20, 1944).

But apparently much time elapsed before Leopold worked on the book. To Hochbaum on December 4, 1944, he wrote, "I'm saying nothing of the essays because I've not yet tackled them." Heavy correspondence continued between them, but no references to the essays are to be found during 1945–1947. A rejection letter from the University of Minnesota Press dated January 31, 1946, suggested that in his essays Leopold "introduce more of himself, so that his personal experience becomes the thread on which the essays are strung."

Knopf continued to check periodically with Leopold on how the work was progressing. Little was being done. At the urging of some of his friends, Leopold sent some of his earlier "philosophical" essays to Knopf. Returning them, Clinton Simpson again expressed concern over what he considered thematic and stylistic disunity. As encouragement he added, "I find whatever you write full of interest and vitality, and it seems to me our only problem is one of fitting together the pieces in a way that will not seem haphazard or annoying to the reader" (April 29, 1946). Said Leopold in reply, "I entirely agree with you that I can see no easy way of getting unity between the philosophical papers and the descriptive essays" (May 10, 1946). No issue was the object of more of Leopold's literary attention than this matter of unifying natural description and ecological exhortation. He saw "Draba" and "The Land Ethic" as of one piece. Knopf did not.

Much writing and revising took place from the last half of 1946 up to the time the manuscript was submitted in the fall of 1947. Leopold's correspondence reveals his determination to get the essays published one way or another. Probably about seven of the forty-one essays eventually published as *A Sand County Almanac* were written during this period.

The results of Leopold's 1946–1947 literary efforts appeared in the manuscript then entitled *Great Possessions,* sent to Knopf on September 5, 1947. It is important to note that Leopold explicitly stated that he put together this manuscript in a deliberate effort to meet the objections Knopf raised in their three letters of July

24, 1944; August 24, 1944; and April 29, 1946. *A Sand County Almanac* as we have it today was Leopold's best effort to combine narrative and exposition, natural fact and conservation value, joy and concern, the particular and the universal, the scientist and the poet and the philosopher.

Knopf saw it otherwise. The book, they said, "is far from being satisfactorily organized. . . . What we like best is the nature observations, and the more objective narratives and essays. We like less the subjective parts—that is, the philosophical reflections which are less fresh, and which one reader finds sometimes 'fatuous.' The ecological argument everyone finds unconvincing; and as in previous drafts, it is not tied up with the rest of the book." As final advice Clinton Simpson suggested to Leopold that "instead of trying to cover so much territory, you might concentrate on the 120 acres of woodland you bought" (November 5, 1947).

By this time Leopold was less ready to accept Knopf's judgment on a matter that had received so much of his attention. Regarding the essays he said in his letter of reply, "I still think that they have a unity as they are" (November 18, 1947).

Leopold did not stop working on the manuscript. Five essays, "Good Oak," "The Geese Return," "Prairie Birthday," "Ave Maria" (better known by its current title "The Choral Copse"), and "Axe-in-Hand" were either written or revised during the fall of 1947 and the winter of 1948. Moreover, the foreword dated July 31, 1947, which had been written for the Knopf manuscript, was replaced with a new foreword dated March 4, 1948, the one with which readers are currently familiar.

During March, Leopold sent his revised manuscript to at least two publishers, Oxford University Press and William Sloane Associates. On April 14, 1948, Oxford University Press accepted the manuscript without critical comment. Sloane was also looking favorably at the manuscript at that time.

It is important to examine what was included in the final manuscript, the one that evolved over six years of debate concerning its contents, the one in which Leopold properly unified the elements of natural description and ecological concern, of field experience and contemplation, of dawn at his Wisconsin River shack and his analysis of environmental history. A comparison of the table of contents of the 1948 manuscript with the 1944 list demonstrates change in two directions. In the first place, by 1948 many more essays were based on the shack experience. Leopold called this section of personal narratives "Sauk County Almanac." In them emerges Leopold the man, the participant observer, "the Standard," as Hochbaum would have it. The arguments of the earlier ecological essays and of the philosophical essays are demonstrated in the personal experiences recorded in the later essays. Fact and value appear in all three essay types: one merely gains access to them through different doors.

Leopold's late attention to essays based on his shack experience is beyond doubt. It is reasonable to assume that the focus grew out of Knopf's and Hochbaum's urging. The 1948 manuscript includes twenty-one essays in Part I, then called "Sauk County Almanac," only eight of which had appeared earlier as articles, two of them, "The Geese Return" and "Pines Above the Snow," in versions so different that one may conclude that fifteen of the twenty-one essays in Part I were then new to the book. Essay manuscript dates and the 1944 list of essays indicate that most of the essays in Part I date from Leopold's last few years and reflect his turn to personal experience and to nature description as a vehicle for conservation thinking. Observed meadow mice replaced criticized road builders.

By contrast, nine of the sixteen essays in Part II had been previously published. Ten of these sixteen essays were among those on the 1944 list. Possibly none of the essays in Part II manuscript were written after mid-1944. Clearly the ecological essays included in Part II reflect Leopold's 1941–1945 sense of what nature writing should be. "Marshland Elegy" and "Thinking Like a Mountain" cover the range of such ecological essays. Any reading which attempts to reconstruct the making of *A Sand County Almanac* must begin with Part II. Parts I and III come later in Leopold's thinking and reflect his final conviction that personal descriptive essays on the one end, experience based ecological essays in the converging middle, and philosophic essays on the other end, all have their place in a book about nature.

It is true that Leopold gave late attention to personal narrative in the shack essays, but it is also true that only late did he decide to include the four philosophical essays of Part III. "Conservation Esthetic" and "Wildlife in American Culture" are only minimally different from their first appearances in 1938 and 1943 respectively. Although "Wilderness" dates from the middle of 1944, Leopold seems not to have seriously considered including it until the 1946–1947 period. "The Land Ethic" contains some new materials as well as substantial portions taken from three earlier published works—"The Conservation Ethic" (1933), "A Biotic View of Land" (1939), and "The Ecological Conscience" (1947), which was first given as a speech in June 1947. Without doubt "The Land Ethic" as we now know it dates no earlier than July 1947. Clearly Leopold considered a concern over the development of a conservation ethic to be no less important in 1947 than he had considered it to be in 1933. Of even more importance for this investigation of the conceptual evolution of *A Sand County Almanac*, Leopold to the end saw a unity in "Draba" and "The Land Ethic." Thus the book in its final form acquired a greater range in style and point of view than it had had in the earlier years of its making. Thematically it remained complexly tight.

Throughout the writing of the book Leopold was concerned about unity of tone, "literary effect" as he described it to Hochbaum. That he was consciously concerned about textual balance is evident from a handwritten, undated sheet entitled "Notes for Paper Writing," in which he argued the need to set forth at the outset the facts and descriptions related to a matter. This process he called "exposition." Only upon completion of this phase ought one move on to what he called "commentary," that is, discussion, appraisal, and interpretation. That organization is certainly reflected in the three-part division of *A Sand County Almanac:* I. "Sand County Almanac"; II. "Sketches Here and There"; III. "The Upshot." Lest there be any doubt, Leopold carefully spelled out their relationship in the March 4, 1948, foreword, which, it seems certain, Leopold wrote with Knopf's previous objections in mind. Part I, he said, tells what his family sees and does at its weekend refuge—personal observation. Part

The Making of *A Sand County Almanac*

II he described as episodes in his life bearing on conservation issues—experientially based conservation commentary. The essays of Part III he said, deal with "philosophical questions" with which "only the very sympathetic reader will wish to wrestle" (viii). Notice how Leopold had been conditioned to assume a limited interested readership for such essays. How wrong Knopf was to want them out; how right Leopold was to insist on their presence. *A Sand County Almanac* is a working out of what "Notes for Paper Writing" says about composition and argument. It begins with the facts and descriptions of land and of man at the shack. Only then does it move on to discussion, appraisal, and interpretation.

The evolution of Leopold's conception of his book, which he never thought of by the publisher's title, *A Sand County Almanac,* is reflected in his changing choice of title. In early 1944 he preferred as title *Marshland Elegy*—a lovely but devastating ecological essay, one in which Leopold does not himself appear. By mid-1944 Leopold considered *Thinking Like a Mountain* a better choice—a more personal and thoughtful essay. The title Leopold chose for the 1947 manuscript was *Great Possessions,* after the essay he thought was his best (October 31, 1944), one which depends on Leopold the man, the phenologist, the lover of land, the man in search for harmony with his world. This last choice reflects Leopold's final and deepest sense of the book: a book which assesses ironically man's great possessions; a book which, through narrative and exposition, both implicitly and explicitly, sets up Aldo Leopold as Standard. As his essays evolved, Leopold added Leopold the example to Leopold the preceptor. Only then did he become Leopold the Standard.

A reading of *A Sand County Almanac* benefits not only from an understanding of its conceptual and structural evolution, but also from an awareness of how Leopold wrote and revised, and of how the book was edited.

Those who have studied the literary process are familiar with the personal journal extracted and reshaped into essays and/or into books. John Muir and Henry David Thoreau wrote that way. Aldo Leopold did not. It was habit for Leopold to take a small pocket notebook with him into the field. The plaid-covered note-book found in his pocket at his death, written in scratched script and cryptic style, contained records of temperatures, shopping lists, correlations of bird songs and candle power, and flower-blooming dates. All is factual and quantitative. The last line in the fire-scorched notebook reads "lilac shoots 2″ long." These field notes were promptly recorded in what is called "The Shack Jour-nal," the several volumes of which are to be found in the Leopold Papers of the University of Wisconsin–Madison Archives.

The journal entries consist entirely of listings and descriptions of natural events, made by different members of the Leopold family, but primarily by Aldo himself. The journal, carefully done and without erasures, is divided into sections with re-curring headings such as "Phenology," "Mammals," "Broken Candle," "First Bloom," "Out of Bloom," "Last Bloom," and so on. This journal, containing only natural description with no value or delight response, is neither a Leopold, nor a family, nor a shack journal. It is a land journal. The Leopold children used it as a data base for later technical articles. It is in no way the basis for the essays in *A Sand County Almanac*. "The Shack Journal" was indexed by Leopold's wife. (Interestingly Leopold had sug-gested to Knopf that *Great Possessions* be indexed.) A search through the journal entries on, for example, woodcocks, demon-strates beyond doubt that the journal is not the literary source of "Sky Dance."

Leopold's initial literary unit was the draft essay, written nei-ther at home nor at the shack but in his office early in the morn-ing. I have found no indication that Leopold wrote from exten-sive preparatory notes. To the contrary, in several drafts I have found blank spaces where exact numbers and technical terms were to be added later. An examination of Leopold's neat pencil drafts reveals writing that is skillful, colorful, natural. The lan-guage of his first drafts was metaphoric, balanced, poetic. There can be no doubt about Leopold's literary gift. Revision followed, mostly deletion and tightening; but occasionally large portions were added, like the Jonathan Carver and John Muir sections of "Bur Oak" and the accounts of geese near the shack in "The Geese Return." Revisions in perspective occur. For example, the

The Making of *A Sand County Almanac*

first draft of "January Thaw" speaks of "the naturalist." The next draft replaces "the naturalist" with "you." Only in the final version does the personalized "I" appear.

Leopold actively sought critical comment on his writing from friends, family, and colleagues. Dozens of people at one time or other examined and criticized Leopold's essays. One of the last notes he wrote before he died requested that eleven of his friends read his manuscript with an eye for unity, style, and balance between description and commentary—matters which concerned him to the end.

After his death, Leopold's son Luna headed a team responsible for the book's final editing. Frances and Frederick Hamerstrom, former students and friends of Leopold and later renowned wildlife biologists in central Wisconsin, wisely argued that the work remain as Leopold left it. Others suggested changes, some of which were made. Oxford considered the title *Great Possessions* unsellable. Worse titles were suggested: *Fast Losing Ground, Last Call, Two Steps Backward, This We Lose,* and others (July 19, 1948). One member of the editorial team, Alfred Etter, found either "Sand Country Almanac" or "Seasons in the Sand Country" a better rubric for Part I than Leopold's choice, "Sauk County Almanac" (June 10, 1948). Exactly how the title *A Sand County Almanac* was chosen remains unclear. Possibly the most important change to the manuscript after Leopold's death was the climax-altering decision to shift "The Land Ethic" from its original first position in Part III to its present final position. Many less important changes were also made. "The Alder Fork" was moved from Part II to Part I. "Ave Maria" was changed a bit and retitled "The Choral Copse" although Leopold's Catholic wife, Estella, preferred the original title. "Prairie Birthday," not in the original manuscript, was added to Part I. "The White Mountain" was renamed "On Top."

Opinions of the essays differed widely. One member of the editorial team thought all of Part I was weak. Bill Vogt considered the essay "Draba" to be "as insignificant as the plant itself." The first sentence of this essay drafted on March 16, 1943, was cut: "During this the longest winter, of this the biggest war, in this month of the big tax, it is salutary to think upon Draba." That the deletion improved the piece seems doubtful. A reference to Gabriel Heater was dropped from "Too Early." Several people on the editorial team objected to language here and there which they considered "too sweet" and not in character for Leopold. But beyond other small changes, the manuscript remained intact.

A Sand County Almanac in the making had a long and interesting history. Any serious study of this work, and certainly any attempt to analyze it as genre exemplar should take into account how Leopold wrote and revised, and especially how the concept of the book evolved. Leopold's attitude toward writing was no less ecological than was his attitude toward land.

Note on the Sources

A. Unpublished Works by Aldo Leopold
   The unpublished letters and essays cited are found in the Leopold Papers (9/25/10), University of Wisconsin–Madison Archives, in the following series and boxes:
Series 1: Correspondence
   Box 1: A–K
Series 2: Organizations and Committees
   Box 3: Delta
Series 6: Writings
   Box 1: Reprints (bound)
   Box 2: Reprints
   Box 5: Sand County Almanac, Correspondence and Drafts
   Box 9: Misc. Mss (pub), Working Papers
   Box 16: Unpub. Mss, AL's [Leopold's] Desk File
   Box 17: Unpub. Mss, Typescript Copies
   Box 18: Early Drafts, Xerox Copies
Series 7: Diaries and Journals
   Box 3: Archery and Shack Journals
B. Published Works by Aldo Leopold
   "A Biotic View of Land," *Journal of Forestry* 37, no. 7 (September 1939): 727–30.

The Making of *A Sand County Almanac*

"Bur Oak Is Badge Of Wisconsin," *Wisconsin Agriculturalist and Farmer* (April 5, 1941): 10.

"Conservation Esthetic," *Bird-Lore* (March–April 1938): 101–9.

"The Ecological Conscience," *Wisconsin Conservation Bulletin* 12, no. 12 (December 1947): 4–7.

"Marshland Elegy," *American Forests* 43, no. 10 (October 1937): 27–29.

"Pines Above the Snow," *Wisconsin Conservation Bulletin* 8, no. 3 (March 1943): 27–29.

"The Plover is Back From Argentine," *Wisconsin Agriculturalist and Farmer* (May 16, 1942): 10.

"Sky Dance of Spring," intended for publication in *Wisconsin Agriculturalist and Farmer* in April 1941, and included in a collection of unpaginated reprints of Leopold's articles from that journal.

"When Geese Return Spring Is Here," *Wisconsin Agriculturalist and Farmer* (April 6, 1940): 18.

"Wildlife in American Culture," *Journal of Wildlife Management* 7, no. 1 (January 1943): 1–6.

5 ∗ Anatomy of a Classic

JOHN TALLMADGE

Compared to other genres in the library of English literature, natural history writing fills only a modest shelf. Yet here too one can find genuine classics such as Gilbert White's *The Natural History of Selborne* (1789), Darwin's *Voyage of the Beagle* (1839), Thoreau's *Walden* (1854), and John Muir's *My First Summer in the Sierra* (1911). Like them, Leopold's *A Sand County Almanac* not only gratifies on first perusal but stimulates and refreshes even after many readings. Like all true classics, it only grows richer with time, and this is no mean accomplishment for an author who thought of himself primarily as a scientist and who had some difficulty in finding a home for his manuscript.

Why, among the less memorable works it so closely resembles, has *A Sand County Almanac* become such a favorite? Leopold has been justly praised for applying ecological principles to land-use planning and for developing the notion of a land ethic. But earlier scientists had introduced the term *ecology* in popular writings well before the turn of the century, and earlier writers, such as Muir and Thoreau, had suggested that humans should view themselves as members of a larger community of life. A student of history might conclude that Leopold merely popularized these ideas, pointing out that his work became widely known only during the late 1960s, in the heyday of environmentalism. But to a student of literature, these facts do not fully explain the book's lasting appeal.

To understand its sense of profound, unfailing richness, I believe we ought to examine *A Sand County Almanac* as a work of

art. As with any piece of fine writing, we readers experience it in
stages: we come with a set of generic expectations; we respond
to the text episode by episode; and we form a sense of the work
as a whole, both at the end of our first reading and whenever we
return for another look. Regarding Leopold's text in this light,
we notice at the very start that he has chosen a form that both
fulfills and challenges our expectations of what a nature book
should be. Then, reading beyond the foreword, we see him en-
gage our interest with a dense but transparent style and a charismatic persona. Finally, when we return to the book for a second
or third reading, we realize that he has been writing in parables
that force us into personal acts of interpretation. The book stays
with us because, in reading it, we experience something analogous to what Leopold experiences in nature itself.

The term *nature writing,* which we would conveniently apply to
A Sand County Almanac, usually means descriptive or narrative
literature that falls between scientific reportage and imaginative fiction. This genre can include the most diverse subject
matter without losing its formal distinctiveness. The critic John
Hildebidle has described nature writing as "informal, inclusive,
intensely local, experiential, eccentric, nativist, and utilitarian,
yet in the end concerned not only with fact but with fundamental spiritual and aesthetic truths."[1] Nature writers rely upon scientific information and their own observations but feel free to
explore these in a broader, more imaginative way than professional science would allow. Traditionally, too, nature writers have
taken their inspiration from particular beloved places. In general,
we might say that whereas scientists use experience and observation to generate "facts" (that is, theories), nature writers use scientific facts to enrich and deepen their readers' experience.

Although the roots of nature writing extend to such antecedent forms as the sketch, the letter, the diary, and the travel book,
we can fix its debut at 1789, when the English clergyman Gilbert
White published *The Natural History of Selborne.* This book,
composed of White's letters to two fellow naturalists, established
many of the standard features of the genre: reliance on scrupulous observation, reverence for fact, an affectionate regard for

other creatures, an informal yet vivid prose style, episodic structure, diversity of material, and a charming local interest. White's
book has gone through countless editions and has been repeatedly invoked as a formal and aesthetic standard by subsequent
writers, including Emerson, Hawthorne, and John Burroughs.
During the century and a half between White's and Leopold's
work, nature writing evolved considerably. The rise of romanticism led to an increased emphasis on the personality of the
naturalist, particularly his emotional responses to the land and
its creatures. Thus Darwin's *Voyage of the Beagle* (like White's
book, a perennial favorite) not only gives meticulous descriptions
of animals and plants but also records its author's outrage at slavery, his awe at the vast deserts of Patagonia, his exhilaration upon
reaching the crest of the Andes, and his increasing wonder at the
immensity of geologic time. Like White, Darwin rearranged and
shaped the raw material of his diaries for publication, but he did
so in order to present his voyage as an idealized circumnavigation where his geographic progress mirrors his developing
understanding of nature.[2] Likewise, John Muir presents his ecstatic experiences in the High Sierra as examples of how people
should relate to the land, and he depicts them as steps in the process whereby he himself was converted to "natural religion."
  Romanticism also gave rise to an American tradition of social
criticism based on the idea of nature's moral purity.[3] Early novelists like James Fenimore Cooper had imagined an American hero
maturing under the virtuous influence of the wilderness, and
painters like Thomas Cole and Asher B. Durand had depicted
sublime landscapes enduring the rise and decay of civilizations.[4]
But Thoreau was the most vigorous practitioner of this sort of
cultural criticism, and his example, like White's in England, set a
standard for all subsequent nature writing in America. Indeed,
as I will suggest later, Leopold seems to have more in common
with Thoreau than with any other nature writer.
  By the mid-twentieth century nature writing was an established genre with a considerable history. No practicing naturalist
would write without a high degree of literary self-consciousness,
and no reader could approach a book like Leopold's without
expectations of what he would find. These facts troubled the

Anatomy of a Classic

publishers to whom Leopold sent his manuscript, as Dennis Ribbens has explained in this volume's previous chapter. In particular, Leopold's correspondence with the editors at Alfred A. Knopf shows that they wanted a book of local observations and personal adventures, with ethical and ecological concepts mentioned only in passing. In other words, they wanted a Gilbert White book. But, as the letters show, Leopold felt that this model would not meet his needs.

Of course, *A Sand County Almanac* does answer many of our initial expectations. Like White, Leopold relies heavily upon his own observations, conveys a strong sense of place, responds with charming affection to the creatures he studies, writes in a vivid, unaffected style, and organizes his book episodically. Like White's romantic successors, Leopold foregrounds his own personality, indulging in social criticism as well as romantic effusions of praise and stressing the moral and aesthetic dimensions of what he sees.

However, Leopold's book exceeds our generic expectations in two important ways. First, though the title suggests otherwise, its scope is not local but continental. Leopold ranges from Sonoran deserts to arctic tundras with stopovers in the basin and range, the midwestern prairies, and the mountains of Arizona and New Mexico. Though the midwest gets the most coverage, all Leopold's descriptions convey an equally strong sense of place. The title of Part II, "Sketches Here and There," and Leopold's claim in the foreword that they are "scattered over the continent," suggest that these pieces are meant to be representative as well as particular. Thus, he addresses continental truths by means of local vignettes. Without sacrificing the strong sense of place, his natural history ceases to be "parochial" in the manner of White or Thoreau and becomes universal.

Second, as we learn from Leopold's foreword, this book will depart from the ostensibly casual and informal modes of presentation used by most earlier nature writers. At the very outset, Leopold makes us conscious of the deliberate artifice with which he has structured his text. The seasonal arrangement of sketches in Part I not only conceals their thematic connections but also reminds us of how carefully some of his predecessors organized their accounts. One thinks, for example, of the idealized year that Thoreau used to structure *Walden* or the idealized circumnavigation Darwin chose for his *Voyage of the Beagle*. Yet Part I more closely resembles White's brand of natural history than do Parts II or III. Part II, in fact, is written in the mode of confessional autobiography and recounts, in Leopold's words, "some of the episodes in my life that taught me, gradually and sometimes painfully, that the company is out of step" (viii). And Part III departs from natural history altogether by dealing with ecological and ethical issues in the abstract; it presents, in the most polemical terms, a program for social transformation. Thus, from the standpoint of form, the book appears to combine three separate genres, and we can understand why Leopold's editors were suspicious: by playing so boldly with his readers' expectations, he risked confusing or alienating them.

The foreword, however, does give a clue to his deeper intentions. The sketches of Part I, written for the most part in the present tense, describe the life he and his family live now. It sounds like an attractive life because it has a purpose (rebuilding what is lost elsewhere) and because it provides spiritual refreshment, "our meat from God" (viii). The sketches bear out these claims, depicting a life of wisdom, excitement, and intimate delight that any reader of nature books might envy. Part II, with its autobiographical episodes (written, for the most part, in the past tense), depicts the transformation of the narrator from an ignorant, insensitive, restless, and ordinary person (someone much like the reader) into the vibrant, perceptive, wise, and serene observer of Part I. Like all conversion stories, Part II inspires emulation. Thus, by the time we reach the abstract and programmatic essays of Part III, we are more ready to entertain their ideas and apply them to our lives than if we had simply read them first. The book's structure seems designed to create a climate of belief that will make us receptive to Leopold's doctrine of land citizenship.

Thus, *A Sand County Almanac* fulfills our generic expectations in some respects while challenging them in others. While it contains much natural history writing in the tradition of White and his successors, its purpose is not merely to delight and instruct.

Like White, Leopold loves the creatures he studies and tries to explain their behavior, but, unlike White, he seeks actively to change his readers. *A Sand County Almanac* is a subversive book. It questions the deepest values of our civilization and challenges us personally on every page. In so doing, it allies itself more closely with Thoreau's method of social criticism based on the standard of nature. Yet Leopold differs from Thoreau by maintaining his role as a professional naturalist. He never lets his moral and social concerns carry him out of sight of the land itself.

When we get beyond the foreword, two features of *A Sand County Almanac* impress us at once: the brevity of its style and the personality of its narrator. From the first sentences of "January," one feels a curious tension in Leopold's prose. On the one hand, it flows smoothly and effortlessly. Each word drops into place with that sense of inevitability that Dylan Thomas said he found in all good poetry. The narrator's manner is confident and relaxed; his tone, though earnest, is rather light and conversational. He prefers a limpid, everyday vocabulary, avoiding jargon, scientific names, and verbal pyrotechnics. His words, in short, do not call attention to themselves. Nevertheless, one senses that each word carries a great deal of meaning, as if chosen with the utmost care. The smoothness and transparency of Leopold's prose belies its density. Like hand-rubbed wood, its surface conceals its craft. Viewed as a whole and compared to other works in the genre, his book achieves an epigrammatic conciseness reminiscent of the sermons of Jesus, the *Tao te-Ching* of Lao-tse, or Chomei's *An Account of My Ten-Foot Square Hut*.

Leopold succeeds here by using what students of short fiction call "techniques of compression." At one level, all storytellers face the same problem: how to present setting, characters, themes, and action in a way that holds the reader's interest and advances the plot. Novelists, blessed with a capacious form, can develop their characters, provide background, paint a scene, or complicate their plots at leisure. But short-story writers must squeeze their narratives into the space of a few pages without stinting on plot, characterization, or theme. Two principal strategies assist them here, and Leopold uses both. The first might be called "concentration": focusing the storyline on the climax and eliminating whatever does not advance the plot. The second technique might be called "engagement," a sort of narrative shorthand whereby the text invites the reader to supply vital information at which it can only hint.

Though Leopold's sketches are presumably factual, they are still good stories and thus have plots, characters, and themes. Like most nature writers, Leopold focuses on his interactions with other creatures, but unlike many of them he presents no fact that does not speak to the point. Whereas for someone like Darwin or White a natural fact may be intrinsically interesting, for Leopold such interest is not enough. The fact must serve the theme and thrust of the essay. Hence, Leopold's sketches, while conveying a good deal of knowledge, seem more coherent and memorable than those of most other naturalists. In his hands, natural facts take on a symbolic richness of implication, and in this respect he resembles both recent writers like John McPhee, a master of "symbolic journalism," and earlier writers like Thoreau, for whom natural facts always suggested higher truths. However, Leopold differs from Thoreau in that he never allows the moral theme to weaken his reverence for fact. Nor does he impose the theme upon the facts, but rather draws it out of them. In brief, he reads human life in the context of nature and not the other way around. Hence he avoids the sentimentalism of the popular writers whom John Burroughs scathingly termed "nature fakers."

Leopold's second technique of compression—what I have called "engagement"—invites the reader to contribute information that the text does not provide, thereby reducing the amount of explanation while increasing the density of implication. Leopold achieves this through repeated use of simple rhetorical figures, notably synecdoche, allusion, irony, understatement, and rhetorical questions. It is these "turns of phrase," rather than a self-consciously poetic vocabulary, that give his prose its memorable succinctness.

Synecdoche—the gesture of letting a part stand for a whole or an individual for a class—operates on the content as well as the form of *A Sand County Almanac*. In "January," for instance, the narrator confesses that his experience arises as a "distraction," suggesting that there is nothing special about it: it was not deliberately sought out, it could have happened to anyone, and it might well have included events other than the ones actually observed. In short, the experience is presented as both specific and representative. The same holds true for the episodes within the story itself. Leopold follows a skunk track, but he might just as well have followed the prints of a deer or a mink. On the way, he is distracted by signs of a field mouse, a hawk, rabbits, and an owl, but other creatures might just as well have appeared. These encounters illustrate general principles of ecology to the naturalist-narrator, whose acts of interpretation (given explicitly in the text) exemplify the sort of attitude we ought to take toward the land. Likewise, *Draba* stands for any "small creature that does a small job quickly and well" (26), the green lagoons of the Colorado represent any shining wilderness of one's youth, and Escudilla illustrates the sort of thing that might happen to any virgin ecosystem invaded by ignorant people. Each particular story invites us to imagine the same thing occurring in other regions. Synecdoche thus enables Leopold to limit the size of his book and restrict the contents of its episodes while at the same time addressing the most profound and universal questions.

Allusion also works to enrich the text, but does so by engaging the reader's knowledge rather than his imagination. Often a few key words will invoke a train of association that significantly affects our interpretation. In "January," for instance, Leopold notes that the skunk track runs "straight across country, as if its maker had hitched his wagon to a star and dropped the reins" (3). This image from Emerson's late essay "Civilization" invokes the Transcendentalist method of viewing nature spiritually.[5] Leopold will pick up that suggestion in "Good Oak," where he mentions the "spiritual dangers in not owning a farm" (6). But the image also tells us how to interpret the rest of this essay. To hitch your wagon to a star, as Emerson implies, is to let the heavens direct your economic life, and this, in a sense, is exactly what the creatures in a land community do. But the skunk also seems to be following some sort of celestial summons, as his odd emergence and abnormally straight track suggest. What does it mean, then, for the naturalist-narrator to follow this skunk? If he meets it, he may also learn something about the higher laws or divine powers that direct it. This winter walk is thus not merely a saunter punctuated with distractions, but a kind of vision quest. It is a quest, however, whose rewards come only from the distractions, since, when Leopold reaches the pile of logs into which the track disappears, he finds neither skunk nor star, but only the tinkle of dripping water that got him out of bed in the first place. The vision is withheld, and the naturalist turns home "still wondering." Yet, though his original questions remain unanswered, he has still found "meat from God" in the form of lessons in ecological interdependency and the perils of self-centered thinking.

Leopold uses another allusion to help us interpret these crucial "distractions." At the end of his encounter with the field mouse, he remarks, "To the mouse, snow means freedom from want and fear." Here he echoes Franklin D. Roosevelt's "Four Freedoms," connecting this apparently trivial incident to the entire context of contemporary politics: the depression, the New Deal, and World War II. We therefore realize that this is a tale not just of mice but also of men. Leopold personifies the mouse as a "sober citizen" and describes his "well-ordered world" in terms appropriate to a human metropolis, thus dignifying the mouse considerably while gently belittling human enterprise. In the end, however, the mouse's fatal self-centeredness stands as a warning to humans, who also place themselves at the center of things. Leopold's text invites us to see each world in terms of the other and thus gain a larger understanding. His allusion to the Four Freedoms challenges us to examine our own priorities and to entertain an ecological interpretation of history. Thus, he accomplishes here in only a few words what he will later take paragraphs to explain in his polemical essay, "The Land Ethic."

A third technique of compression uses rhetorical figures that invite the reader to say what's left unsaid. Irony, understatement, and rhetorical questions occur throughout the text. For instance, at the end of his ecstatic recollections of the Colorado delta, Leopold writes, "All this was far away and long ago. I am told the green lagoons now raise cantaloupes. If so, they should not lack flavor" (148). It is ironic, of course, that human beings would destroy an earthly paradise to grow breakfast foods, and Leopold's comment must be one of the bitterest in all literature. Similarly, in "Illinois Bus Ride," when Leopold passes a chanelled streambed, he remarks dryly, "The water must be confused by so much advice" (119). Further down the road he comes to a spanking-new farmstead where "even the pigs look solvent" (119). However, the fields are plowed right up to the fence, and the creek shows signs of flooding and gullying. "Just who is solvent?" he asks. "For how long?" (119). The alert reader quickly imagines the answers: not the pigs (who will soon be slaughtered) and not the farmer (whose soil will soon be eroded), but only the land itself, which is already being dissolved and carried away.

These techniques of compression both abbreviate and deepen Leopold's text, resulting in a peculiarly attractive style. Because the narrator does not browbeat us with verbiage, we feel respected, as if he valued our time, and so we are more inclined to listen. Here, we feel, is a writer who has taken pains to find exactly the right words to express the distilled wisdom of his life. Such "verbal behavior" gives us important insights into his character and helps create the vivid narrative persona which so compels our attention on a first reading.

In any narrative, the reader must come to terms with the personality, values, and apparent intentions of the "implied author." This intimate relationship creates special problems for writers of confessional autobiography, who seek to justify their mature beliefs. They must affirm the views to which they have been converted, yet they must also convince their readers that the conversion was genuine, and that means helping their readers identify with their younger selves. Similarly, nature writers, who already love, understand, and appreciate other creatures, must win the sympathy and trust of their less sensitive and less well-informed readers. In both cases, the narrators seek to win their readers' trust and inspire emulation while at the same time challenging their normal behaviors and fundamental beliefs.

Leopold's naturalist-narrator gains our confidence in several ways. To begin with, he salts the text with references to his university, his graduate students, the jargon and nomenclature of his discipline, and the systematic field studies he has undertaken. Unlike White, Thoreau, and Muir, he bands chickadees and subjects migrating geese to statistical analysis: in short, he's a professional. But, like his predecessors, he is also well versed in literature, history, and philosophy, as we learn from his numerous allusions. He also relies heavily upon his own observations. Indeed, he is as scrupulous and sensitive as White, with the same contagious curiosity, the same interest in field observation, and the same knack for making connections. Finally, as an accomplished hunter and outdoorsman, he seems to feel an emotional bond with his subjects different from what we find in Muir or Thoreau. The fact that he has killed woodcock, for instance, helps make the praise of "Sky Dance" not only more poignant but more profound. The essay is tinged with a darkness we have not encountered before, and this creates an impression that its wisdom is somehow more genuine.

But Leopold's narrator appears more than professional to us: we also find him an entertaining, engaging, and charismatic personality. We note at once his affectionate and sentimental regard for other creatures, an attitude quite reminiscent of White, Darwin, and Muir. Leopold loves Canada geese, sandhill cranes, and wild game of all kinds, yet he seems charmingly fond of small creatures that no one appreciates, like *Draba,* the chickadee, and the field mouse. His interest in undervalued and marginal things extends to landscape; he prefers the nondescript scenery of a sand county farm to the romantic sublimities of Muir's High Sierra or the edenic woods and pastures of White or Thoreau. Indeed, landscape as such hardly seems to interest him. What goes on *in* the land is what fascinates, and toward this he reveals an endearing capacity for the deepest feelings. His essays on wild geese, for instance, convey as much yearning, exultation, and praise as the most strenuous poem of Shelley. Leopold's sense of beauty is complex, as J. Baird Callicott explains elsewhere in this volume, but it is romantic and involving rather than abstract and contemplative. Thus, it contributes to our sense of him as a warm and engaging person: we admire and are drawn to thinking people who can also be deeply moved.

Two other attractive features of Leopold's persona deserve special mention. He appears to us as a man whose convictions run as deep as his feelings. His long intimacy with the land seems to have bred both loyalty and a sense of moral responsibility. He considers it his duty to defend the land by bearing witness to the truth of which he has been convinced. The certainty with which he states his views, and the courage he shows in casting them before an uncaring public, create an aura of charismatic self-confidence. At the same time, he avoids melodramatic posturing by poking fun at himself from time to time. Thus, in "Good Oak," when the family goes back to sleep after a lightning strike, he remarks, "Man brings all things to the test of himself, and this is notably true of lightning" (8). And, in "Red Lanterns," he exchanges roles with his dog, whom he imagines treating him, the professional naturalist, as just another "dull pupil" (64).

Finally, Leopold's persona challenges or confronts us, even as it draws us out and wins our trust. This naturalist-narrator presents himself as an outsider, devoted to an unpopular cause yet convinced that "the company is out of step" (viii). Time and again he calls our attention to the glaring disparity between our view of the land and how the land actually behaves. At times he reacts with bitter sarcasm, as when, in "Good Oak," he quotes a Wisconsin governor's declaration that "'state forestry is not a good business proposition'" (10). Leopold does not call the governor a fool; he merely notes that "it did not occur to him that while the courts were writing one definition of goodness in the law books, fires were writing quite another on the face of the land. Perhaps, to be a governor, one must be free from doubt on such matters" (11). We can hardly help wincing for this poor gentleman. Leopold's dry wit reminds us of the opening chapter of *Walden*, where Thoreau adopts the same crusty manner (rather like that of a yankee farmer) to rouse his neighbors from their lives of quiet desperation. In both cases we, as readers, shrink from the object of scorn. Leopold's barbs entertain, but they also challenge us to examine our values. We would hardly like to have such a critic turn his wit upon *us*.

But this clever, accomplished, and formidable narrator is not always a happy one, and here again he challenges us. A deep current of melancholy runs through this book, even in the midst of its most rapturous celebrations. It soon appears that the narrator's conversion experiences have not really made his life any easier, but instead have made it more complicated and painful. To love a place is to suffer doubly when it perishes: witness Leopold's reaction to the loss of the green lagoons. To be able to read the book of nature means being able to see the ugliness and degradation as well as the beauty. Only Leopold notices the yellow cheat grass taking over the slopes of California foothills; only he knows why the cheat has replaced the more desirable native grasses. The price of his ecological wisdom is loneliness, isolation, and an aching sense of loss. If we wish to emulate him, we have to accept those terms.

To sum up, we may conclude that the brevity of Leopold's style reflects the values and personality of his narrator. Both serve the purpose announced in the foreword: to change the way readers conceive of and respond to the land. We might say that Leopold presents himself as a prophet, someone with special knowledge, a history of transformative experiences, and a "strange power of speech." Like the Old Testament prophets, Leopold finds truth in the wilderness and comes back to warn a society with little sense of its own spiritual danger. Like the New Testament prophets, he finds no honor in his own country, which certainly does not wish to be changed. Therefore, he resorts to the only weapons prophets have ever been able to wield: the strength of truth and the transforming powers of language. He takes his place with Thoreau as an American Jeremiah, judging his culture against the standard of wild nature.

Before we have gone very far in *A Sand County Almanac,* Leopold's sketches and arguments begin to make a cumulative impression. We finish the work convinced of its unity and purpose, and, when we return to it, we bring a clear sense of its moral vision. As with a great work of fiction, our knowledge of the ending only increases our delight in the storyteller's art. Here, understanding the doctrine of the land ethic only makes Leopold's vignettes more resonant.

Anatomy of a Classic

A Sand County Almanac has more thematic unity than many classics of natural history. White's *Natural History of Selborne* may demonstrate a method, but it hardly advances a philosophical or moral position; what unity it achieves comes from the personality of the narrator and the narrow range of his excursions. Similarly, Darwin's *Voyage of the Beagle* gathers the most diverse researches and concerns under the umbrella of an idealized chronology and the viewpoint of a single narrator; it has, in this sense, little more unity than a picaresque novel. *A Sand County Almanac,* in contrast, presents its sketches as moments of insight that point toward a core of truth the way iron filings respond to a hidden magnet. Only upon rereading the book do we fully appreciate how much its stories and underlying themes reinforce each other. Eventually, we realize that Leopold's "sketches" are really parables and that his parabolic style accounts more than anything else for the book's perennial freshness.

A parable conveys novel ideas by means of familiar facts and situations, as in the well-known parables of Jesus. The story of the mustard seed and the tares (Matthew 13) uses agricultural imagery to teach celestial truths. Jesus explains that the kingdom of heaven is like "the least of all seeds," yet when it is grown, "it becometh a tree, so that the birds of the air come and lodge in the branches thereof." Likewise, he compares God to a farmer whose enemy sows tares among his wheat, thinking to ruin the harvest; but the farmer waits till the crop is full grown and then has his servants bind the tares and cast them into the fire. Both scenarios must have been quite familiar to Jesus' audience, yet to interpret the parables one needs a rudimentary sense of Christian doctrine, as we learn a few verses later when Jesus' dull disciples ask for an explanation. To interpret, one must first believe, or at least entertain the possibility of belief. Then, the act of interpretation can give the doctrine new meaning through application to our life. But this application familiarizes the doctrine and so increases our incentive to believe. Thus, the parable initiates a spiraling process of interpretation and belief (sometimes referred to as the "hermeneutic circle"). Presenting a truth in this way, rather than by argument, has the advantage of winning from the reader not only understanding but commitment.

The "doctrine" behind *A Sand County Almanac* might be said

to rest on three central ideas, which Leopold states in his fore-word. These are that land is a community, that land can yield a cultural harvest, and that land should be loved and respected. These ideas are conveyed by three archetypal images occurring throughout the text: the analogy between natural and human systems (such as economics or politics), the metaphor of nature as a language or system of signs (the "book of nature"), and the figure of the naturalist-narrator as an example of how humans ought to relate to land. All three emerge clearly in the first two essays, which thus provide a key for reading the rest of the book.

In "January" the narrator's main discovery is that both natural and human communities are governed by similar economic con-cerns. The narrator shows enough love and respect for the skunk to follow his track, leave him alone in the driftwood pile, and write a charming essay about him later. In "Good Oak" the nar-rator's main concern is reading the book of nature, which he does by examining the traces of recent activity (rabbits, present seedlings, squirrels planting acorns) and the history of the land as recorded in the chronicle of the oak tree itself. Reading the book of nature is thus both investigative ("reading sign") and in-terpretive (reconstructing the past). The naturalist expresses his love and respect not only by referring to the oak in honorific terms ("dead veteran," "no respecter of persons," "what is a ton of ice, more or less, to a good oak?") but also by ritualistically returning its ashes to the soil so that they may rise to new life in another form.

To illustrate how the parabolic style works, let me conclude with a look at "Thinking Like a Mountain." Thanks to Dennis Ribbens, we know that Leopold once considered this essay for his title piece, and certainly it is one of his most memorable. The title, at first glance, appears cryptic and paradoxical, much like a Zen koan: how could a huge, inanimate object "think"? The an-swer does not emerge till the end, and, like the solution to a koan, it requires a fundamental change in the reader's world view. The essay points the way for this change by presenting a story of conversion and inviting us to identify with the narrator. In this sense, it reflects the ultimate concern of the book as a whole.

"Thinking Like a Mountain" begins with a vivid evocation of a wolf's howl. Imagined, not recounted, and put in the present tense, this opening gives the reader an immediate vicarious experience. The next two paragraphs state the theme: how to "decipher the hidden meaning" of that call, or, by extension, how to read the signs in the book of nature. Paragraph two presents a series of misreadings, much like those of the creatures in "January" who misunderstand why snow falls or grass grows. Only the mountain, it seems, can interpret the call "objectively," avoiding the fatal errors committed by self-centered beings with a limited sense of history. By the end of the third paragraph, then, we already sense the moral of the conversion story we are about to hear: that reading the book of nature will yield a "deeper meaning" and put us in touch, through a shared point of view, with the mountain itself.

Paragraphs four through six shift to the past tense, presenting the experience that converted the narrator to his present practice of thinking like a mountain. It is a brief, vivid, and intensely dramatic tale with a violent climax and a mournful, repentant ending. As he watches the green fire die in the old wolf's eyes, the narrator feels his triumph turn to a horrified realization of sin. But his sin is explicitly described as an act of misinterpretation: "I realized then, and have known ever since, that there was something new to me in those eyes" (130). Up to this point, the narrator's understanding has been as limited and subjective as those of the creatures he lists in the second paragraph. But he regrets it now: "I was young then, and full of trigger-itch" (130). His retrospective view comes with the bitterness of tragic insight. He now recognizes the consequences of misreading nature's signs: the murder of a fellow-creature, and his own alienation from a greater sentient being (the mountain itself).

The second section of the essay, divided from the first by asterisks, describes later experiences that confirm the truth glimpsed at the moment of conversion. Here the narrator bears witness to the worth of his new belief. Significantly, the encounter with the dying wolf has empowered him to read the book of nature correctly. He notices the slopes of other mountains "wrinkle with a maze of new deer trails" (130) once the wolves have been killed.

He can explain the bleached deer bones and denuded shrubs of the dying range. But he can also read the history of the land and predict its future, because he can understand the rules by which the signs he reads are produced. That is, he can think ecologically; he can think like a mountain. In the second paragraph we learn that the term "mountain" means not just a pile of rock but the entire community of creatures who live there. "Mountain" is thus synonymous with "range," and "thinking like a mountain" means considering the needs of the whole community rather than those of one member alone. The third paragraph extends this insight to cows and ranchers, warning that the latter's own chronic misperception will one day prove fatal: "Hence we have dustbowls, and rivers washing the future into the sea" (132). This final declaration shows that the narrator's conversion experience and subsequent correct interpretations have turned him into a prophet, thus fulfilling the role he assumed in the foreword.

The third section of the essay presents what might be called the moral. Here the narrator steps back from his experiences and predictions, discarding the intense voices of penitent and prophet to assume a serene and reflective tone. He ponders the deeper connection of ecological principles and events to our own moral and social life. His scornful allusion to Neville Chamberlain's slogan "peace in our time" suggests a link between incorrect interpretation and moral laxity: we deliberately misread because we want to feel safe. But too much safety, as the starved deer know, may lead to death in the end. The essay concludes by invoking the wisdom of Thoreau, our foremost prophet-naturalist, but in such a way that we must construe his words from a new ecological viewpoint. The final sentence returns to the image of the wolf call with which we began and once more raises the question of "hidden meaning." Though we cannot be sure of its message— as the narrator's "perhaps" reminds us—we are a lot wiser (and sadder) than we were at the outset. Now the title no longer seems paradoxical: now we can imagine what it *could* mean to "think like a mountain." Thus the essay comes full circle, beginning and ending with the problem of interpretation. Its moral, I would suggest, is that everything we thought we understood— the wolf, game management, the raising of livestock, our laws, our politics, even the words of our prophets—must now be reviewed from an ecological perspective. Thus, the structure of the essay reflects the parabolic movement of interpretation itself, and, in so acting to transform our consciousness, it serves as a paradigm for the effect of the book as a whole.

A Sand County Almanac has become a classic because of the ways it arouses and holds our interest: by playing upon our expectations of what a nature book should be, by creating a dense but transparent style and an attractive but challenging narrative persona, and by conveying its wisdom in parables that change our angle of vision. To call it a classic, however, is not to say it is perfect, merely that it endures. It rewards rereading with increased delight and deeper, more personal instruction. Like Keats's Grecian urn, it works to tease us out of thought and into imagination, not by virtue of its contents, but by the manner in which they are conveyed. That Leopold had a rare literary gift cannot be doubted, and one wonders what more he might have written if he had lived longer. But what can one add to the distilled wisdom of a lifetime? The gospels, too, are brief, challenging, and wonderfully durable. We could ask the same question of them: not, what more do we want, but what else do we need?

Notes

1. John Hildebidle, *Thoreau, A Naturalist's Liberty* (Cambridge: Harvard University Press, 1983), 61.

2. See John Tallmadge, "From Chronicle to Quest: The Shaping of Darwin's *Voyage of the Beagle*," *Victorian Studies* 23 (Spring 1980): 325–46.

3. See Perry Miller, "The Romantic Dilemma in American Nationalism and the Concept of Nature," in *Nature's Nation* (Cambridge: Belknap Press of Harvard University Press, 1967), 197–207.

4. See Miller, "Romantic Dilemma," 197–98; also Barbara Novak, *Nature and Culture* (New York: Oxford University Press, 1980).

5. Ralph Waldo Emerson, "Civilization," in *Society and Solitude* (Boston: Houghton Mifflin, 1904), 30.

6 ✻ The Conflicts of Ecological Conscience

PETER A. FRITZELL

Few books do more than *A Sand County Almanac*, in themselves and in the history of their publication and reception, to illustrate the complexities of ecology as it appeared on the American scene in the late 1960s and early 1970s. Few books have satisfied the conceptions of ecology of more readers than *Sand County* has. Few books have come so close to containing the several meanings of ecology that have become current in the late twentieth century—ecology as science, ecology as subject matter, ecology as ethical and aesthetic point of view, and even ecology as preferred environment.

Ecology and the ecology movement, however, are hardly sufficient to explain the popularity and reasonable longevity of *Sand County,* and certainly not to explain the substantial coterie—by now almost a subculture—of historians, philosophers, legislators, citizen scientists, lawyers, environmental activists and educators, social commentators, landscape planners, architects, and even literary critics, who have devoted themselves to quoting, analyzing, summarizing, refining, and even amplifying its major terms and conceptions. As the essays and attendant bibliographies in this centenary collection make abundantly clear,

The original version of this essay, here revised and shortened, was published in *Transactions of the Wisconsin Academy of Sciences, Arts and Letters* 64 (1976): 22–46.

The Conflicts of Ecological Conscience

there is much more to *Sand County* than meets the usual popular eye—much more, in fact, than its many casual readers have ever suspected—enough, surely, to lead a discerning reader to think that we may be dealing with an American classic, a classic of the kind that captures the basic drama of a culture.

Whatever else it may be, *Sand County* is a quintessentially American book, quintessentially American because, in its twentieth-century vocabulary, in its dedication to the science of ecology, it dramatizes—however subtly and unobtrusively—the traditional competing allegiances of historic America. In so doing, the book proves the continuing vitality of that paradoxical American determination, on the one hand, to reform or redeem history—in the process, restoring or reclaiming an original harmony between human and nonhuman nature—and, on the other hand, to escape history entirely, by turning (or returning) to nature, as we say, by letting nature take its course, by leaving nature to its own devices, or by simply appreciating nature and its workings in all their deep evolutionary power and ecological complexity, wherever they may lead.

Like other American classics—like Cooper's *The Pioneers,* or *Moby-Dick,* or Faulkner's *The Bear—Sand County* is defined in part by a deep, abiding commitment to a programmatic—scientific, ethical, and finally social—solution to the problems and disharmonies of mankind's historic relations to nonhuman nature. Like *The Great Gatsby, Walden,* and Cather's *My Antonia, Sand County* is devoted in part to that perennially Western and especially American dream of perpetual harmony among self, society, and nonhuman surroundings. At the same time, and like each of those other American classics, *Sand County* is bound to and by a conviction, just as deeply held and at least as abiding, that, finally, historic time—the time of programmatic solutions and ethical axioms—is not the time that matters, that all human endeavors and all human history must be held and kept in long-term evolutionary and ecological perspective, that all human history ends in paradox rather than solution or resolution. *Sand County,* then, is a surprisingly complex and intricate work, rewarding to anyone who would closely examine its overt argument, its many covert questions and counterarguments, the relationships among them, and the methods used to present them.

The primary argument of *Sand County*, the argument for which Aldo Leopold is famous, is a twofold statement: a descriptive illustration and explanation of land communities—what they are, how they work, and how they change—and a closely related, prescriptive declaration of needs served by maintaining certain kinds of land communities.

As the book develops, its primary argument moves logically, from the land community as empirical fact, through the recognition of man's place in land communities, to a plea for ethical standards of land use.* The argument progresses inductively: from one restricted land community; through a set of more loosely structured and less detailed land communities; to a discussion of the concept of land community—from detailed description and narration of a single land community; through description, narration, and exposition of several diverse land communities; to a largely expository discourse on the aesthetics, ethics, science, and culture of land communities. In essence, the relation of *Sand County*'s Part I, "A Sand County Almanac," to Part II, "Sketches Here and There," to Part III, "The Upshot," is the relation of percept to generalized observation to concept.

Part I establishes the land community as an empirical (i.e., descriptive and narrative) certainty. It presents a series of essentially mundane facts in the life of a Wisconsin landowner. It speaks of pasque-flowers, geese, chickadees, mice, grouse, deer, cornfields, high waters, and old boards, among other things. It is noticeably lacking in conceptual terms. Conceptions and concepts, when they appear, are colored, qualified, and finally overshadowed by perceptual terms. The whole is broken into what might be called perceptual situations. Even individual chapters are occasionally fragmented. "October," for example, begins and

* Throughout this essay I have frequently used the term *man* and the masculine pronoun *he* as synonyms for mankind. Where I have done so, I have done so because such usage has seemed truer to the text and the narrating figure of *Sand County* than other available alternatives.

The Conflicts of Ecological Conscience

ends with grouse hunting but presents, in between, a deer, a chickadee, some geese, some ducks, and a marsh. "December" begins with a canine rabbit hunt and then jumps to more chickadees, deer tracks, grouse, pine trees, and finally to chickadee 65290. Through such a perceptual conglomeration, the members of a land community are introduced, and along with them another significant member of the community—a man perceiving, digesting, and pondering a set of basic materials and relationships in a restricted environment.

Of first importance to the methods and meanings of Part I are the perceptual raw materials that form the substance of the man's surroundings. Without meadow mice, old boards, and chickadees, the man would amount to very little. Meadow mice, grouse, and deer tracks substantiate the man's experience, not to say his identity. Pine trees, high waters, and woodcock corroborate the existence of the land community. But the man and his reflections are also central to the environment, as they are to the primary argument of the book as a whole. Without the man and his reflections, neither meadow mice, nor old boards, nor chickadees would amount to much.

Part I is made of more than perceptual raw materials, however crucial they may be to the major statement of *Sand County*. The simple sense experiences of the man are occasionally crossed and complicated by symbolic reflections and interpretive analogies. Trout, as they rise to the man's brown miller and eventually land in his creel, call to his mind a similar human disposition—perhaps even his own—"to seize upon" gilded morsels containing hooks (39). Grouse that thunder across narrow openings in tamarack swamps suggest to the man that "many thoughts, like flying grouse, leave no trace of their passing, but some leave clues that outlast the decades" (57). And the long growth of pines in 1941 leads him to wonder whether these pines "saw the shadow of things to come" and "made a special effort to show the world that pines still know where they are going, even though men do not" (83).

Sometimes the man in Sand County reads the book of nature rather heavily. At other times he only suggests symbolic intricacies to a perceptual situation. In "January," for example, a meadow mouse "darts damply" across a skunk track, leading the man to wonder:

> Why is he abroad in daylight? Probably because he feels grieved about the thaw. Today his maze of secret tunnels, laboriously chewed through the matted grass under the snow, are tunnels no more, but only paths exposed to public view and ridicule. Indeed the thawing sun has mocked the basic premises of the microtine economic system!
>
> The mouse is a sober citizen who knows that grass grows in order that mice may store it as underground haystacks, and that snow falls in order that mice may build subways from stack to stack: supply, demand, and transport all neatly organized. To the mouse, snow means freedom from want and fear. (4)

In situations like this one, the basic perceptual substance of the land community momentarily recedes into the background, as the narrator becomes a symbolist, in this case perhaps a satiric symbolist. The mouse becomes an analogue for the narrator's conception of economic man, and mouse tunnels become metaphors for the inroads civilized man makes on the land. A midwinter thaw may be equivalent to the passage of time that exposes man's economic determinism. It may be that man, like the mouse, is a sober citizen who will continue to believe that grass grows to make haystacks (which may be fed to cattle, which may profitably be sold to other men)—"supply, demand, and transport all neatly organized." Sometimes the man in *Sand County* reads the book of nature so suggestively that his readings can only be said to have multiple meanings. At still other times, he does not seem to read the book of nature at all, but simply to present it, without interpretation.

As Part I develops, the Sand County land community and the personality of the man in that community develop coordinately. Through his symbolic interpretations of seemingly mundane events, he becomes more than a recorder of details or a personifier of plants and animals. With the meadow mouse he becomes a socioeconomic critic of sorts. When he makes wood in "February" he becomes a historiographer—saw, wedge, and axe, in turn, becoming three distinct, if complementary, approaches to the past. When he interprets the "December" pine (which has its own

The Conflicts of Ecological Conscience

"constitution" prescribing terms of office for its needles) he becomes yet another kind of ironist, a commentator on the relations of human language and the so-called nonhuman environment.

The episodes of Part I have the quality of developed perceptions. Their denotative and connotative impression is cumulative rather than progressive. The prose is basically descriptive, narrative, and dramatic (rather than expository or imperative). The voice, the point of view, is fundamentally personal rather than collective or impersonal.

If Part I of *Sand County* validates the land community, Part II extends that validation, taking it beyond personal experience and carrying it across conventional geobiotic and cultural boundaries. Where Part I concentrates attention on a single psychobiotic locus, Part II covers several loci in a much broader field of reference. Where, in Part I, explanations are the dramatized thoughts of narrating or narrative character, in Part II they are also (and often) rendered independent of specific narrative occasions. The narrative "I," "my," and "me" of Part I often become, in Part II, expository "we," "our," and "us"; or generic "you" and "your." The largely psychobiotic drama of Sand County becomes, in substantial part, the sociobiotic exposition of Wisconsin, Illinois, Iowa, Arizona, New Mexico, Chihuahua, Sonora, Oregon, Utah, and Manitoba. The largely personal and local history of Sand County tends to become generic—regional, American, and even Western. In more senses than one, "Sketches Here and There" is an expansion of "A Sand County Almanac."

In fact, the second part of *Sand County* is a hybrid of the styles that define the first and third parts, a stylistic amalgam of the concrete and the abstract, personal narrative and impersonal exposition, idiosyncratic perception and generalized conception. As it extends the style of "A Sand County Almanac" it also leads into "The Upshot."

In "Manitoba," for example, one reads not only a past-tense personal narrative about grebe-watching, but also an impersonal, present-tense interpretation of grebes and grebe-watchers:

> I was starting to doze in the sun when there emerged from the open pool a wild red eye, glaring from the head of a bird. Finding all quiet, the silver body emerged: big as a goose, with the lines of a slim tor-

pedo. Before I was aware of when or whence, a second grebe was there, and on her broad back rode two pearly-silver young, neatly enclosed in a corral of humped-up wings. All rounded a bend before I recovered my breath. And now I heard the bell, clear and derisive, behind the curtain of the reeds.

A sense of history should be the most precious gift of science and of the arts, but I suspect that the grebe, who has neither, knows more history than we do. . . . If the race of men were as old as the race of grebes, we might better grasp the import of his call. Think what traditions, prides, disdains, and wisdoms even a few self-conscious generations bring to us! What pride of continuity, then, impels this bird, who was a grebe eons before there was a man. (160–61)

Specific, narrative grebes become the archetypal grebe. The events of personal, narrative experience are rendered exemplary and set in expository and collective context. Past-tense personal narrative leads to self-reflection, and reflection leads to what "we" characteristically do, to what "we" might be, to a shared human condition. Personal narrative is explained and subsumed by impersonal exposition.

Of the six chapters in Part II, only the short chapter "Illinois and Iowa" maintains the unbroken personal narrative prose of Part I. In the other five, personal experience of and in land communities is rendered collective, generically human, and increasingly abstract. The chapter "Wisconsin" is almost entirely discursive and impersonal. Even its most personal and narrative segment, "Flambeau," ends in a brief historical account of the REA, the Conservation Commission, and the Legislature. In "Arizona and New Mexico" the nonhuman environment is identified primarily in its relations to "Homo texanus," and the narrator becomes a horseman, an "undistinguished" member of a sociohistorical and human community of cowmen, sheepmen, foresters, and trappers. "Oregon and Utah" is dedicated to an explanation of cheat grass and its effects on the American West, an explanation interrupted only once by a personal narrative that illustrates and corroborates prior, impersonal exposition.

As personal experience of land communities is generalized, so, of course, are the detailed events of the geobiotic environment. As the first-person singular gives way to the first-person plural (or the second-person), so it also gives way to the even less personal third-person:

> High horns, low horns, silence, and finally a pandemonium of trumpets, rattles, croaks, and cries that almost shakes the bog with its nearness, but without yet disclosing whence it comes. At last a glint of sun reveals the approach of a great echelon of birds. On motionless wing they emerge from the lifting mists, sweep a final arc of sky, and settle in clangorous descending spirals to their feeding grounds. A new day has begun on the crane marsh. (95)

The descent of sandhill cranes in "Wisconsin" gives rise to a discussion of their historicity; to notes and comments on a Holy Roman Emperor, Marco Polo, and Kublai Khan; and, finally, to historiographical ponderings: "Thus always does history, whether of marsh or market place, end in paradox" (101).

The stylistic strategy of Part II of *Sand County* is to take the details of a man's relations to land communities, to generalize them, and lead them toward the major concepts and arguments of "The Upshot"; to gradually withdraw the personal voice, turning ever more frequently to the materials of history and community, and increasingly to outright socioeconomic criticism.

Arguments at best only implicit in Part I become increasingly explicit in Part II. Judgments at best tentative in Part I become increasingly overt: "That the good life on any river may likewise depend on the perception of its music, and the preservation of some music to perceive, is a form of doubt not yet entertained by science" (154). As personal experience is generalized, as social and economic events become primary subjects of concern, so the apparently unassuming personal observations and reflections of Part I tend to become discrimination and adjudication. Prescriptive terms, such as "overgrazing" and "misuse," multiply. Cheat grass is "inferior," and research is a "process of dismemberment."

As adjudication increases, so, appropriately, does ratiocination—the formation and explanation of conceptions necessary to support normative judgment. The notion of the land pyramid, for example, becomes in Part II something more than the unnamed thought of a Sand County landowner, and yet something less than the "mental image" it will be in "The Upshot":

Food is the continuum in the Song of the Gavilan. I mean, of course, not only your food, but food for the oak which feeds the buck who feeds the cougar who dies under an oak and goes back into acorns for his erstwhile prey. This is one of many food cycles starting from and returning to oaks, for the oak also feeds the jay who feeds the goshawk who named your river, the bear whose grease made your gravy, the quail who taught you a lesson in botany, and the turkey who daily gives you the slip. And the common end of all is to help the headwater trickles of the Gavilan split one more grain of soil off the broad hulk of the Sierra Madre to make another oak. (152–53)

As the details of personal experience are rendered collective and abstract, as prescription and explanation take over from description and narration, so the unsifted percepts of Part I are gradually built into conceptions, conceptions that will become concepts in Part III.

If Part I of *Sand County* is about things like a meadow mouse, a Wisconsin landowner, and chickadee 65290; and Part II about things like horsemen, government trappers, and sandhill cranes; Part III is about things like wilderness, recreation, science, wildlife, conscience, aesthetics, conservation, ethics, land health, the A-B cleavage, and the community concept. "The land pyramid"—a complex of oak, buck, cougar, and goshawk in Part II—becomes in Part III a "symbol of land," an "image," and "a figure of speech":

> Plants absorb energy from the sun. This energy flows through a circuit called the biota, which may be represented by a pyramid consisting of layers. The bottom layer is the soil. A plant layer rests on the soil, an insect layer on the plants, a bird and rodent layer on the insects, and so on up through various animal groups to the apex layer, which consists of the larger carnivores.
>
> The species of a layer are alike not in where they came from, or in what they look like, but rather in what they eat. Each successive layer depends on those below it for food and often for other services, and each in turn furnishes food and services to those above. Proceeding upward, each successive layer decreases in numerical abundance. Thus, for every carnivore there are hundreds of his prey, thousands of their prey, millions of insects, uncountable plants. The pyramidal form of the system reflects this numerical progression from apex to

base. Man shares an intermediate layer with the bears, raccoons, and squirrels which eat both meat and vegetables. (215)

The oaks, jays, and bucks of Part II become more abstract plants, birds, and animals—logical components of "the biotic pyramid" rather than characteristic members of regional ecosystems. The individualized actors of Part I—the dog, the meadow mouse, and the landowner—give way to carnivores, herbivores, and mankind.

In more ways than one, Part III is the upshot to *Sand County*. In formal terms, it is the ideational conclusion to a logical and stylistic order that moves inductively from the narrative raw materials of Part I through the generalized observations of Part II. As the land community of Part I and the regional communities of Part II become the concept of land community in Part III, so the landowner of Part I and the community member of part II become the ethicist and moralist of Part III. Thoughts, impressions, and preferences that are functions of a first-person narrative and narrating character in Part I become theoretical constructs in Part III. Judgments and criticisms that are expressions of shared experiences in Part II become in Part III the reasoned end products of a formal normative system, a moral code for man's relations to nonhuman environments.

As "The Upshot" to *Sand County* develops, "mental images" and concepts become primary subjects of concern. Figures of speech—"the land community" and "the land pyramid"—become the philosophic cornerposts to a land ethic. Symbols—"the biotic pyramid" and "the pyramid of life"—become necessary psychosocial conditions to developing an ecological conscience. In short, the primary argument of *Sand County* is made explicit—the argument from ecosystem as fact and concept to the need for maintaining certain kinds of ecosystems.

In one sense, it is an easy argument to follow. Its descriptive and nomothetic elements are easy to understand, and its prescriptive or normative components seem to grow logically from systematic premises and historical evidence: Land is, and for a long time has been, a complex organism, a "highly organized structure" of interlocking food chains and energy circuits. The continuous functioning of land depends, and for a long time has depended, on "the co-operation and competition of its diverse parts," of which man is simply one among many. "The trend of evolution is to elaborate and diversify the biota." "Evolution has added layer after layer, link after link," to the pyramid of life; and man is but "one of thousands of accretions" to its height and complexity. Man, however, has often behaved as if he were an overlord rather than a citizen of the land community. Modern man especially has simplified (or oversimplified) the land pyramid. He has, in fact, been a counterevolutionary force in the biota. He has had counterevolutionary effects on the environments he has occupied. He has depleted soils and deranged the circuits of energy-flow that sustain the land. He has upset the capacity of land for self-renewal. He has thought of land as property and of himself as property owner. He has applied to land a narrow system of strictly economic priorities and values. He has thought of himself as possessing the land rather than being possessed by it. As a result, both he and the land are in need of a new system of concepts and values; a system that will assure the continued existence of empirical norms for healthy land through the preservation, conservation, and restoration of lands that have not suffered the most disruptive inroads of civilization; a system of values and images that will restore, and then maintain, harmonious relationships between man and land.

To many people it is a satisfying, if not compelling, argument—the argument from land community as fact and concept to the land community as value. It derives not only from the intimate narrative experiences and personal preferences of Part I but also from the generalized observations and collective experiences of Part II. At the same time, it contains or encompasses its logical components. It explains the relations of man to land, both the relations of the individual man as they appear in "A Sand County Alamanac," and the relations of historic, human communities as they are expressed in "Sketches Here and There." It calls for "an internal change in our intellectual emphasis, loyalties, affections, and convictions"; an "extension of the social conscience from people to land." And it rests finally on the proposition that *Homo sapiens* must begin to think of itself as "plain

The Conflicts of Ecological Conscience

member and citizen" of the land community rather than as "conqueror" of it: "A thing is right when it tends to preserve the integrity, stability, and beauty of the biotic community. It is wrong when it tends otherwise" (210, 209, 204, 224–25).

Compelling and satisfying though it may be, direct though it may appear, the primary argument of *Sand County* is far more complex than any simple summary of its development can suggest—in part because its descriptive and prescriptive components are at stylistic and conceptual odds with one another; in part because the dialectic of their relationship is typically covert; and in part because that dialectic changes form as point of view and prose style change.

Like virtually all nature writing—and much of classical American literature—*Sand County* is dedicated to and defined by two mutually exclusive conceptions of man's relations to nature: one basically descriptive, synthetic, and holistic; the other essentially prescriptive, analytic, and dualistic. Man in *Sand County* is, and ought to be, a plain member and citizen of the land community. But he is also an exploiter and subverter of land communities, and ought not to be. He is, whether he likes it or not, an overseer and guiding force in the biotic community, "a King . . . one / Of the time-tested few that leave the world, / When they are gone, not the same place it was" (223). Yet he is also but "one of thousands of accretions" to the pyramid of life, and cannot be otherwise. Nature, analogously, is the self-sustaining system of energy circuits that contains and absorbs all humans and their artifacts. But it is also that which humans naturally attempt (and must attempt) to contain and absorb, if not in supermarkets and power plants then in "mental images" and symbols. The land pyramid is a "mental image" in terms of which humans must conceive their actions. Yet it is also the contextual system, the "revolving fund of life," in which all those actions are taken, including actions leading to the creation of "mental images."

To be a part, yet to be apart; to be a part of the land community, yet to *view* or *see* one's self as a part of that community (and, thus, to remain apart from it)—that is the dilemma. And these are the classical desires—to be a part of the biotic pyramid, yet to

know the pyramid and the terms of one's position in it; to identify man in terms of his environment, yet to know the terms of
that environment and the terms of man's place in it; to present
the land pyramid as an accurate description of man's relations to
the environment, yet to present the land pyramid as a "symbol"
for land, a symbolic key to an ethical system created and held by
humans, and not very many humans at that. Both conceptions
are as conventional in nature writing, both as traditional in classic American literature, as they are definitive in *Sand County*. So,
too, are the impulses they express and the needs they seek to sat-
isfy. The one—a holistic conception of man's place in nature—
aspires to a non-normative theory of the development and operation of the geobiotic environment, a disinterested account of
the relations of organisms (including humans) and their surroundings. The other—a dualistic conception—aims at an at
least partially normative theory of man and nature; a bilateral,
and at least partially adjudicative account of man's relations to
the geobiotic environment; a conception of man and nature
based on fundamental distinctions between the natural (i.e.,
geobiotic and, therefore, appropriate) actions of man and at least
some of his civilized (i.e., social and economic) habits.

The holistic conception of man's place in nature is dedicated to
the proposition that human behavior—however distinctive,
however cultural, however linguistic—is finally, and fully, ex-
plainable in the same basic terms as the behavior of other organisms. The holistic perspective draws no fundamental distinctions between natural and civilized human actions. It identifies
man as an integral part of the land community, and other members of that community as integral parts of man and man's environments. It explains the actions of man, whatever their form, as
functions in and of ecosystems. So, too, it explains ecosystems
(and the behavior of their constituents) as functions in and of human communities. It represents evolution as a process subsuming
human history and containing man, even as man foreshortens
food chains and "deranges" the "normal" succession of nonhuman ecosystems. It emphasizes an integral, ongoing connection
of man and land, even as man "destroys" land, even as changing
lands provoke changes in man—the whole to be traced through

time. And it, therefore, expresses, and no doubt satisfies, that deep human desire to be immersed in one's surroundings.

The dualistic view of man and nature, by contrast, presupposes that some of man's relations to land are integral, and some are not; that some members of the land community have integral relations to man, and some do not. Resting on the proposition that "man-made changes" in the biotic pyramid "are of a different order than evolutionary changes," it represents evolution as a process in which humans participate only imperfectly, a process in which they may early have participated but now, very often, do not. Taking essentially nonhuman biotic communities as norms, it explains human actions as functions in and of evolving ecosystems only when those actions are consonant with the needs of other elements in such systems, where *consonant* means conducive to the continued, healthy existence of all present species—as defined and determined by humans and human science. The dualistic perspective, then, identifies man as an integral part of the land community only as human actions perpetuate and sustain that community's component food chains and energy circuits. At the same time, this perspective identifies man as *Homo sapiens,* as a knowing creature capable of altering or directing the course of evolution, a creature whose behavior can only be partially explained in geobiotic or ecological terms. As a complex of ideas and impulses, the dualistic perspective is, thus, biosocial rather than geobiotic, disjunctive rather than conjunctive. It assumes that humans can do (and have done) inimitable things to the pyramid of life, but it also assumes that humans have the capacity (unique among organisms on earth) to rectify their misdeeds, to become (or once again become) plain members and citizens of the land community—only this time knowing, self-conscious citizens. And it, therefore, effectively expresses, and perhaps satisfies, that classical and continuing need in humans (or at least Western and American humans) to be on top of their environments, to transcend (or at least to comprehend) their surroundings and their conditions.

Taken together, these two conceptions of man's relations to nature are not only the logical antipodes to the world of *Sand County;* they are also its warp and woof, its constant stylistic threads. They intersect each other on almost every page of every chapter, and they make the book as a whole a composition of opposites, a fabric of coordinates converging from two radically different directions, a fabric of ironies, ambiguities, and paradoxes. Together they account not only for the frontside of *Sand County*—the overt argument from land community to land ethic—but also for its backside—the covert pattern of questions, doubts, and contrary impulses that runs just behind the primary surface and upon which its overt statements depend.

In Part III, of course, the dialectic cloth of *Sand County,* the web of relationships between holist and dualist, is more abstract and more open than it is in either of the first two parts. In two or three short pages of "The Land Pyramid," for example, one hears both holist and critical dualist: Man is "one of thousands of accretions to the height and complexity" of the pyramid of life, and "the trend of evolution is to elaborate and diversify the biota." In short, man is a plain member and citizen of the evolving land community. "Evolution is a long series of self-induced changes" in the circuit of life, "the net result of which has been to elaborate the flow mechanism and to lengthen the circuit." And yet "man's invention of tools has enabled him to make changes of unprecedented violence, rapidity, and scope" in the biotic pyramid. He has simplified its flow mechanisms and shortened its circuits. His agriculture, industry, and transportation have produced an "almost world-wide display of disorganization in the land," a disorganization that "seems to be similar to disease in an animal, except that it never culminates in complete disorganization or death" (219).

Loosely interwoven as they are in Part III, the perspectives of the monist and the dualist are comparatively easy to separate, and the contradictions between them are readily apparent and inescapable. If man *is* a plain member and citizen of the land community, one of thousands of accretions to the pyramid of life, then he *cannot* be a nonmember or conqueror of it; and his actions (like the actions of other organisms) *cannot* but express and affect his position within the pyramid of life. If the trend of evolution *is* "to elaborate and diversify the biota," and man is an inextricable part of the process, then man *cannot* be simplifying its flow mechanisms or shortening its circuits. If evolution *is* "a long series of self-induced changes" in the circuit of life, and man's actions are inseparable parts of evolution, then "man's invention of tools" cannot logically be said to have enabled *man* to make changes of "unprecedented violence, rapidity, and scope" in that circuit. Conversely, if man's technology *has* enabled man to make unprecedented changes in the circuit of life, then evolution is *not* simply a long series of self-induced changes in that circuit. It is in recent earth history, at least in part, a series of man-induced changes. If man *is* simplifying the flow mechanisms and shortening the circuits of the biotic pyramid, then the trend of evolution is *not* to elaborate and diversify the biota, at least not so long as man is a functioning member of it. If man is an exploiter and conqueror of the land community, then he is not a plain member and citizen of it, or at least he is a citizen only part of the time.

Because its composition is bold, direct, and expository—because its alternative conceptions and arguments are unmediated by narrative occasions or shared experiences—Part III of *Sand County* raises almost as many questions as it seems to answer: Is man a plain member and citizen of the land community? Or is he a conqueror and exploiter of land communities? Or is he both? Is man a citizen of the land community only part of the time? If so, when? Under what conditions? Is he a citizen of the land community when he *thinks* of himself as such, when he consciously seeks to understand his place in the biotic pyramid? Do man's thoughts and language take him outside the land community? Or are his thoughts, conceptions, and ethics (like his search for shelter, food, and sex) simply expressions of his place in the pyramid of life? And, if so, can any fundamental distinction be drawn between his "land ethic" and any other ethic he may apply to land? Is evolution a long series of self-induced changes in the circuit of life? Or is it also man-induced? And, if so, to what extent, when, and under what conditions? If at least some man-made changes in the land are of a different order than evolutionary changes, how is *man* to tell whether or not such a change (say the adoption of a land ethic) is evolutionary?

Questions such as these, lying just behind the surface of "The

Upshot," produce a series of critical uncertainties for the serious reader of *Sand County*. Deriving, as they do, from the clash of disinterested science and interested criticism, they create a pattern of critical doubts and ambiguities, a pattern that surrounds almost every important statement in the final, abstract section of the book. If, for example, "an ethic, ecologically, is a limitation on freedom of action in the struggle for existence," then an ethic (any ethic) is very much like a water supply, a windstorm, or a wheat field; like money, cancer, or language (202). Are any of man's ethics more than expressions of man's geobiotic condition? Have they ever been? Can they ever be? In what sense is a "land ethic" or an "ecological conscience" more than an ecological or evolutionary event, to be understood (as are other such "ethics" and "consciences") as another in the series of "successive excursions from a single starting-point, to which man returns again and again to organize yet another search for a durable scale of values" (200)? Does all history consist of "successive excursions from a single starting-point," successive searches after a durable scale of values? Or is history progressive? Can man find in the land ethic a *final,* durable scale of values? *Have* we learned "that the conqueror role is eventually self-defeating," because "the conqueror knows, *ex cathedra,* just what makes the community clock tick, just what and who is valuable, and what and who is worthless, in community life" (204)? Can, or should, we learn?—"A thing is right when it tends to preserve the integrity, stability, and beauty of the biotic community. It is wrong when it tends otherwise." Is man to determine when the biotic community is stable and beautiful? Or must man take counsel from other citizens of the community—not only pines, deer, and wolves but cheat grass, algae, gypsy moths, and rats? Can man take anything more than *human* counsel with the other members of the land community? Can such counsel ever express more than the ecological interests of humans and the species they most closely identify with? Is the problem *we* face simply a matter of extending "social conscience from people to land"? Are *we* willing to extend to other members of the land pyramid the conscience and the consciousness that would make the notion of land community a working analogy? Or would that simply be

The Conflicts of Ecological Conscience

another human imposition on the pyramid of life, another example of exploitive anthropocentrism?

Virtually every key word in "The Upshot" has two mutually exclusive meanings—one descriptive, the other prescriptive. "Evolution," for example, is the process of change that occurs over time in the geobiotic environment. But "evolution" is also the process by which the land sustains itself, the purpose of which is to preserve the life of *the* biotic pyramid. "Ecological situations," similarly, are networks of organisms and environments changing over time. But "ecological situations" are also the kinds of situations that men ought to seek, the kinds of relations among organisms and environments that must not be violated and which evolution is designed to foster. "The land pyramid" is both fact and value. So are "the pyramid of life," "the land community," and even "the land."

The conflicts between fact and value in Part III of *Sand County* are both radical and unconditional, more radical and less conditional than they are in either Parts I or II. Neither shared regional experiences nor personal narrative occasions are present to relieve or moderate the tension between them. "The biotic pyramid" both "is" and "ought to be." An "ecological conscience" involves both a conscious understanding of the biotic pyramid, a cosmogonic sense of what it is and how it changes, *and* a desire to discriminate its healthy and unhealthy states, a teleological need to indicate where it ought or ought not to go: "In all of these cleavages, we see repeated the same basic paradoxes: man the conqueror *versus* man the biotic citizen; science the sharpener of his sword *versus* science the searchlight on his universe; land the slave and servant *versus* land the collective organism" (223). But is not man a conqueror when he thinks of himself as conqueror? Or even when he *writes* of himself as a plain citizen? Is not science (*scientia*) the sharpener of his sword even when he *styles* it a searchlight? Is not land a slave and servant even when, or perhaps especially when, humans call it a collective organism? Is not man indeed a king, "one of the time-tested few that leave the world . . . not the same place it was" (223)? Does any organism leave the world the same place that it was?

Logical and philosophical questions arise easily in "The Up-shot" to *Sand County*. One might even say "The Upshot" is de-signed to raise such questions—by alternating conceptions of man and nature, by juxtaposing competing theories of history, by rotating "is" and "ought," by interlacing fact and value. At the same time, however, "The Upshot" exposes the basic threads of the book as a whole, the elements that make up its imperative primary surface as well as its interrogative subsurface, in Parts I and II no less than in Part III.

Though they are less obvious in either of the first two parts than in the third, the ambiguities and uncertainties that underlie "The Upshot" are no less central to "Sketches Here and There" or "A Sand County Almanac." In both Parts I and II, the fabric woven of holism and dualism, fact and value, NATURE and na-ture, is tighter than it is in Part III (though considerably more open in the second than in the first).

In "Wisconsin" of Part II, "a new day has begun on the crane marsh":

> A sense of time lies thick and heavy on such a place. Yearly since the ice age it has awakened each spring to the clangor of cranes. The peat layers that comprise the bog are laid down in the basin of an ancient lake. The cranes stand, as it were, upon the sodden pages of their own history. These peats are the compressed remains of the mosses that clogged the pools, of the tamaracks that spread over the moss, of the cranes that bugled over the tamaracks since the retreat of the ice sheet. An endless caravan of generations has built of its own bones this bridge into the future, this habitat where the oncoming host again may live and breed and die.
>
> To what end? Out on the bog a crane, gulping some luckless frog, springs his ungainly hulk into the air and flails the morning sun with mighty wings. The tamaracks re-echo with his bugled certitude. He seems to know.
>
> * * *
>
> Our ability to perceive quality in nature begins, as in art, with the pretty. It expands through successive stages of the beautiful to values as yet uncaptured by language. The quality of cranes lies, I think, in this higher gamut, as yet beyond the reach of words. (96)

This passage, like so many others in *Sand County,* presents a divided picture of the natural world. On the one hand, it suggests that man is not a part of nature, that nature—the crane marsh and the events that make it up—is an essentially non-human phenomenon, a set of processes that man participates in only vicariously, however much he may wish otherwise. On the other hand, it also suggests quite the opposite, that man is indeed a part of nature, that nature—insofar as it is *known* and *appreciated*—is at least as human as it is nonhuman, at least as much the product of human ingenuity as it is the conditioner of man's "creative" impulses—an expression of his science, his language, and his needs for order as much as it is their underlying substance.

Perhaps the greater part of the passage implies that nature is foreign to humankind, that nature is never more than inadequately understood by humans. "A sense of time lies thick and heavy" on crane marshes, as it typically does not on human farms and cities. "The cranes stand, as it were, upon the sodden pages of their own history." Humans, by implied contrast, often seem to stand on the pages of a history not their own, their own history being, too frequently, thin and dry. A time, not human time, is the time of the crane marsh. And while humans may, in one sense, know that time—know that lake, mosses, tamaracks, cranes, and peat have built the crane marsh, "this bridge into the future, this habitat"—humans do not know, perhaps cannot know, "to what end," however much they may wish to. The crane, on the other hand, flailing "the morning sun with mighty wings" and bugling his certitude, "seems to know"—not only where he has come from but also where he and his marshes are going—a quality of knowledge man can perceive perhaps, but which he cannot capture in language.

In such a world (at least half the world of *Sand County*), man is a stranger to nature, a questing perceiver of natural processes, an outside observer attempting with little success to encompass and comprehend cranes, crane marshes, and their relations. As outside observer, man only learns slowly to perceive quality in nature. His efforts to capture such quality in language are a never-ending, and seldom successful, struggle to reconcile his own needs and his own terms with the nonhuman world around him. His dilemmas are not the crane's dilemma. He tries to write and understand, while the crane simply goes on living. There is a world of difference between cranes and the man who seeks to know them.

Tellingly, ironically, and inevitably, man's desires to know—his needs to order, explain, and understand (to the extent they are realized)—set him apart from the very things he would know. In his questing, ordering hands, a complex of sounds and silence, cries and mists, arcs and spirals, becomes a crane marsh. The crane marsh, in turn, becomes a product of ecological succession—mosses, tamaracks, cranes, and peat, "all neatly organized"—and more than that, even. For the differences between man and marsh are apparent not only in his scientific propositions but also in his "poetic" figures of speech. Ecological succession becomes "an endless caravan of generations" building futuristic bridges, and the crane in his habitat becomes a phoenix, what some humans call a mythic being, with a capacity for self-renewal and a certitude (if not a determination) that man can only envy.

In man's hands, the crane becomes considerably more than a plain member and citizen of the land community, more than a crane perhaps. For to be a crane in man's ordering hands is not just to be named. It is to be compared with other named things. It is to become a member of complex systems—energetic, genetic, morphological, and ecological—systems in which the thing you are swallowing (what humans call a frog) is no longer a primary term, systems in which "frogs" are replaced by "heat," "waste," "structure," "energy," and "time." To be a "crane" is to be vested with man's hopes and doubts, with man's particular kinds of order.

Strangers though they may be in one sense, crane and man are, in another and no less significant sense, not strangers at all, but acquaintances of the most intimate kind. As stylized products of ecological succession and evolutionary change, cranes and crane marshes express man's needs to know even as they pattern man's knowledge. The crane marsh—the "bridge into the future," the "habitat" for the "oncoming host"—is a method for coming to terms with living, breeding, and dying—for man no less than for the crane. The crane, in turn—the bugling phoenix—is an assurance that life is self-renewing, a means to knowing or imagining that something or someone can answer the question "to what end?" even if man cannot.

Despite apparent differences in their respective media, perhaps the crane's dilemma *is* man's dilemma. Still, it is no doubt only in man's power to say in words, while trying to capture the quality of cranes in words, that "the quality of cranes lies . . . as yet beyond the reach of words." It is no doubt only in man's power to conclude with paradox and yet, paradoxically, continue to seek resolutions to paradox—to say in quite civilized words, in sentences far from "wilderness incarnate,"

> Thus always does history, whether of marsh or market place, end in paradox. The ultimate value in these marshes is wildness, and the crane is wildness incarnate. But all conservation of wildness is self-defeating, for to cherish we must see and fondle, and when enough have seen and fondled, there is no wilderness left to cherish. (101)

In "Sketches Here and There," as one might expect, logico-philosophical problems—ethical, metaphysical, and even scientific questions—are raised in context of regional economics, national politics, and cultural traditions. History and evolution are embedded in regional development and ecosystemic change, in the details of cranes and crane marshes, or coyotes and abandoned logging camps. Man's relations to nature are the crane-watcher's relations to cranes, or the government trapper's relations to the mountain Escudilla. By the same token, the crane-watcher's inability to capture the quality of cranes in words is less an epistemological dilemma than it is a shared, cultural difficulty. The paradoxes of wilderness preservation are less logical problems than they are communal concerns. And problems generated by competing ideas of conservation are less theoretical difficulties than they are "our" problems—emblematic problems that express "our" needs, national problems that "we" have created, regional problems that "we" must solve, if any solutions are to be found.

As one returns from "The Upshot" to "Sketches Here and There," the dialectic cloth of *Sand County* becomes, in one sense, more dramatic and familiar. As farmers and cornfields replace concepts and symbols, philosophic doubts become geohistorical ironies, logical dilemmas become biocultural ambiguities. The basic elements of *Sand County* are held constant, while the patterns they form vary.

As one moves, in turn, from "Sketches Here and There" back to "A Sand County Almanac," the fabric of *Sand County* is further compressed, its dialectic threads are even more closely interwoven than they are in Part II. As the voice of collective experience—regional and historical experience—becomes a personal voice, the voice of the Sand County landowner, so sociobiotic ironies become psychobiotic uncertainties. As the historical time and space of geocultural regions become the personal narrative time and space of a Sand County farm, so logical dilemmas become psychological dilemmas, and philosophical problems become personal problems. Alternative notions of man and nature are absorbed in personal narrative. Competing conceptions of history and evolution are embedded in autobiographical experience. What had been "our" traditions and desires become "my" personal habits and needs, and "our" disagreement becomes "my" uncertainty. What ought to be the case is what "I" wish for; what is the case is what "I" see, and need to see; and any differences between the two are facets of "my" personality.

In "A Sand County Almanac" the dialectic elements of the book as a whole are fully dramatized. Both cultural traditions and philosophic questions are functions of an individual man's relations to his land:

> I find it disconcerting to analyze, *ex post facto,* the reasons behind my own axe-in-hand decisions. I find, first of all, that not all trees are created free and equal. Where a white pine and a red birch are crowding each other, I have an *a priori* bias; I always cut the birch to favor the pine. Why?
>
> Well, first of all, I planted the pine with my shovel, whereas the birch crawled in under the fence and planted itself. My bias is thus to some extent paternal, but this cannot be the whole story, for if the pine were a natural seedling like the birch, I would value it even more. So I must dig deeper for the logic, if any, behind my bias.
>
> The birch is an abundant tree in my township and becoming more so, whereas pine is scarce and becoming scarcer; perhaps my bias is for the underdog. But what would I do if my farm were further

The Conflicts of Ecological Conscience

north, where pine is abundant and red birch is scarce? I confess I don't know. My farm is here.

The pine will live for a century, the birch for half that; do I fear that my signature will fade? My neighbors have planted no pines but all have many birches; am I snobbish about having a woodlot of distinction? The pine stays green all winter, the birch punches the clock in October; do I favor the tree that, like myself, braves the winter wind? The pine will shelter a grouse but the birch will feed him; do I consider bed more important than board? The pine will ultimately bring ten dollars a thousand, the birch two dollars; have I an eye on the bank? All of these possible reasons for my bias seem to carry some weight, but none of them carries very much.

So I try again, and here perhaps is something; under this pine will ultimately grow a trailing arbutus, an Indian pipe, a pyrola, or a twin flower, whereas under the birch a bottle gentian is about the best to be hoped for. In this pine a pileated woodpecker will ultimately chisel out a nest; in the birch a hairy will have to suffice. In this pine the wind will sing for me in April, at which time the birch is only rattling naked twigs. These possible reasons for my bias carry weight, but why? Does the pine stimulate my imagination and my hopes more deeply than the birch does? If so, is the difference in the trees, or in me? (68–70)

"Is the difference in the trees, or in me?"—with that question the Sand County landowner gives the dialectic of *Sand County* as a whole perhaps its purest expression. On a November day, he poses the question implicit in virtually all the logico-philosophical dilemmas and sociobiotic inconsistencies of "The Upshot" and "Sketches Here and There." Is man a plain member and citizen of the land community? Or is he its conqueror? Or is he both? Why do I find man as plain member more attractive than man as conqueror? Does the notion of man as biotic citizen stimulate my imagination more than the notion of man as conqueror? If so, is the difference in man's actions, or is it in me and my notions? Am I a citizen, or conqueror, or both? Is the biotic pyramid a fact? Or is it a figure of thought and value? Is the difference between fact and figure, or fact and value, a function of things in the pyramid of life? Or is it a function of needs in man, and in me? Is there a difference between what man knows and what the crane on the Wisconsin marsh knows? And, if so, is the difference in what each knows, or in me? Are man and nature both inextricable parts of a unified natural whole? Or are man and nature distinct? And, if so, are the distinctions in man and nature, in man, or in me? Crucial though such questions (and their answers) are to the formulation of a land ethic in "The Upshot"; central though they are to the conception of regions in "Sketches Here and There"; in "A Sand County Almanac" they concentrate in one fleeting, reflective, November moment of a self-conscious land-owner's life; they come back to earth, as it were, to the relations of an organism and its environment, the personal relations of a man and his surroundings.

In the landowner's almanac of Part I, the dialectic threads of *Sand County* produce an autobiographical cloth, a closely woven pattern of personal perceptions, individualized arbitrations, and self-conscious reflections. As one might expect, in Part I of *Sand County* epistemological and metaphysical dilemmas become matters of momentary self-interrogation, passing rhetorical queries of a man "wasting" his November weekends "axe-in-hand." Alternative approaches to history become idiosyncratic analogies for "saw, wedge, and axe" as the man makes wood in February. The events of history—a federal law prohibiting spring duck shooting, for example—become personalized analogues for the growth-rings on the oak he is cutting. Members of the human community—neighbors, tourists, and speeding grouse hunters—become the substance of occasional, and often self-gratifying, thoughts. And nonhuman elements of the geobiotic environment become configurations of singularly personal ideas and impressions.

A meadow mouse darts damply across a skunk track, provoking questions and reflections, complex figures of speech and developed conceptions, perhaps even concepts and mental images:

Why is he abroad in daylight? Probably because he feels grieved about the thaw. Today his maze of secret tunnels, laboriously chewed through the matted grass under the snow, are tunnels no more, but only paths exposed to public view and ridicule. Indeed the thawing sun has mocked the basic premises of the microtine economic system! (4)

The Conflicts of Ecological Conscience

Perhaps the mouse *does* suggest man in his current relationships to land, and perhaps some thaw will expose man's habits of land use:

> The mouse is a sober citizen who knows that grass grows in order that mice may store it as underground haystacks, and that snow falls in order that mice may build subways from stack to stack: supply, demand, and transport all neatly organized. (4)

But there is also a strong possibility that the mouse and his tunnels represent each member of the land community, each member of the land community (humankind included) soberly and unconsciously pursuing the mouse-eat-grass, hawk-eat-mouse pattern that prevails in all ecosystems: "To the mouse, snow means freedom from want and fear." To the hawk, in the next paragraph, ". . . a thaw means freedom from want and fear" (4).

Perhaps the mouse, the hawk, and every other member of the land community (man included) will continue to see snows, thaws, and bioeconomic organizations as meaning freedom from want and fear. Maybe it is natural for man, mouse, and hawk to use their surroundings to be free from want and fear. Perhaps it is necessary that man, mouse, and hawk attack and exploit other members of the land community—if not with underground haystacks and economic systems, then with scientific explanations and ethical judgments, with language and figures of speech. In short, just as there is evidence to support an ironic and satiric reading of the meadow mouse episode, so there is evidence to suggest that the episode is nothing more or less than a *picture* of the actions and habits of *one diminutive member* of the land community. Or, to put it another way, just as there is evidence for a prescriptive reading of the situation, so there is evidence to support a descriptive reading.

Such are the ways of analogies and analogues that both the figure and the thing figured are brought to the same end. Only the maker of figures, perhaps, is provided momentary freedom from want and fear, and even that kind of freedom seems terribly fleeting to the self-conscious ecologist. In the end, then, we return to the man in Sand County, to a man in a land community.

III * The Upshot

7 ★ The Land Aesthetic

J. BAIRD CALLICOTT

"These essays deal with the ethics and esthetics of land" (Aldo Leopold, 1947 Foreword [to *Great Possessions*]).

Aldo Leopold's original contribution to environmental *ethics* is universally acknowledged and appreciated.[1] His essay "The Land Ethic" in *A Sand County Almanac,* though to one degree or another anticipated by Thoreau, Darwin, and Muir, is the first self-conscious, sustained, and systematic attempt in modern Western literature to develop an ethical theory which would include the whole of terrestrial nature and terrestrial nature *as a whole* within the purview of morals.[2]

An equally original and revolutionary "land aesthetic" reposes in *A Sand County Almanac.* Taking *Sand County* as a whole, an appropriate aesthetic response to nature seems quite as important to the author as an appropriate ethical attitude. In the foreword, Leopold expressly mentions the "esthetic harvest" we may "reap" from land, as one among the "three [major] concepts" which "these essays attempt to weld" (viii–ix). (The others are the community concept of ecology, and ethics.) And in the oft-quoted summary moral maxim of "The Land Ethic,"

Earlier versions of this essay appeared in *Environmental Review* 7 (1983): 345–58 and in *Orion* 3 (Summer 1984): 16–23. The author expresses appreciation to Allen Carlson for valuable suggestions for improving it for this volume and, more generally, for his extensive scholarship in natural aesthetic theory, upon which the author has freely drawn in preparing this revision.

the beauty of the biotic community is an important measure of the rightness or wrongness of actions: "Examine each question in terms of what is ethically *and esthetically* right. . . . A thing is right when it tends to preserve the integrity, stability, *and beauty* of the biotic community. It is wrong when it tends otherwise" (224–25, emphasis added).

Contemplating human action in respect to land, one might suppose that utilitarian desiderata tempered and occasionally opposed by ethical criteria are preeminently efficacious while aesthetical considerations carry less weight. Historically, however, many more of our conservation and preservation decisions have been motivated by beauty than by duty. Environmental aesthetics is about the beauty of the natural environment. Sound critical thinking upon environmental aesthetics, therefore, may be of even greater pragmatic importance than the formulation of an environmental ethic. An ethic is onerous, burdensome—according to Leopold's own definition, "a limitation on freedom of action in the struggle for existence" (202). Beauty, on the other hand, is attractive. Plato, in the *Symposium,* exposed the link between beauty and eros. Duty is demanding—often something to shirk; beauty is seductive—something to love and cherish. Hence, to cultivate in the public, as Leopold urged, "a refined taste in natural objects," is vital to enlightened democratic land-use decisions.[3] *A Sand County Almanac* aims to do just that, quite as much as it aims to instill an ecological conscience.

The land aesthetic is more diffuse than the land ethic; it is scattered throughout *Sand County* and some of Leopold's other essays (especially those in *Round River*) and never labeled as such. (The "Conservation Esthetic" is, to be sure, included in *Sand County's* "The Upshot," but it is more concerned with the niceties of outdoor recreation than with principles of natural aesthetic judgment.) Here I assemble and juxtapose the scattered fragments of the land aesthetic and abstract from them a systematic theory of natural beauty and the criteria for its appreciation. One cannot, however, begin to comprehend the originality of Leopold's contribution to the appreciation of natural beauty without some knowledge of the tradition which he had to overcome.

The Land Aesthetic

Natural aesthetics is a pitifully underworked topic in Western philosophical and critical literature.[4] From Aristotle to Nelson Goodman, a lot has been written on the subject of aesthetics, but most of it has centered on art—painting, sculpture, architecture, drama, literature, dance, music, and more recently, cinema. Indeed, *aesthetics* and *art criticism* are practically synonymous terms.

Upon review of the history of Western aesthetic *experience* of nature, the absence of a rich Western philosophical literature—ancient, modern, and contemporary—in natural aesthetics becomes less surprising. In sharp contrast to Far Eastern cultures, in the West an aesthetic response to nature actually appears to be a lately acquired taste. Indeed it seems that in Western civilization prior to the seventeenth century, nature was simply not a source of aesthetic experience. Given the nearly universal susceptibility to natural beauty ambient today in Western culture, this realization is monumental, even shocking.

The discovery in the West of the existence of beauty in nature was, as Christopher Hussey points out, ancillary to the representation of nature in the exciting new genre of painting for which the name "landscape" was coined: "It was not until Englishmen [as well as other Westerners] became familiar with the landscapes of Claude Lorraine and Salvatore Rosa, Ruysdael and Hobbema, that they were able to receive any [aesthetical] pleasure from their [natural] surroundings." Hence, cultivated Europeans began "viewing and criticizing nature as if it were an infinite series of more or less well composed subjects for painting."[5]

Natural beauty thus shone forth but, like the moon, by a borrowed light.

The "picturesque" aesthetic, as the name suggests, self-consciously canonized as beautiful those natural "scenes" or "landscapes" suitable as motifs for pictures. It was formulated in William Gilpin's *Three Essays on Picturesque Beauty* first published in 1792 and Uvedale Price's *Essay on the Picturesque,* first published in 1794. The aesthetic analysis of landscapes (i.e., countryside) was thus largely cast in terms of the colors, tones, textures, relative size, and arrangement or "composition" of topographical masses like mountains, valleys, lakes, copses, meadows, fields, streams, and so on.

Christopher Hussey argues that from the mid-eighteenth to the mid-nineteenth century all the arts—painting, poetry, the novel, architecture, even music—coalesced around the picturesque aesthetic. It also spawned two new popular activities: aesthetic travel (or rustic tourism) and the aesthetic management of nature, revealingly called, then as now, "landscape gardening" (in England) or "landscape architecture" (in the United States).

While painting and the other arts moved on to other fashions—romanticism, impressionism, cubism, abstract expressionism, etc.—and artifactual aesthetic theory in the work of John Ruskin, Clive Bell, Roger Fry, George Santayana, and others kept abreast, popular taste in nature remained more or less tied to the picturesque.[6] And natural aesthetic theory languished. Hence we continue to admire and preserve primarily "landscapes," "scenery," and "views" according to essentially eighteenth century standards of taste inherited from Gilpin, Price, and their contemporaries.

The prevailing natural aesthetic thus is not autonomous, but derivative from art; it is conventionalized, not well informed by the ecological and evolutionary revolutions in natural history; and it is sensational and self-referential, not genuinely oriented to nature on nature's own terms. In a word, it is trivial.

Naturally occurring scenic or picturesque "landscapes" are regarded, like the art they imitate, to be precious cultural resources and are stored, accordingly, in "museums" (the national parks) or private "collections" (the landscaped estates of the wealthy). They are visited and admired by patrons just like their originals deposited in the actual museums in urban centers. Nonscenic, nonpicturesque nonlandscapes are aesthetic nonresources and thus become available for less exalted uses. While land must be used, it is well within our means to save representative nonscenic, nonpicturesque nonlandscapes—swamps and bogs, dunes, scrub, prairie, bottoms, flats, deserts, and so on—as aesthetic amenities, just as we preserve intact representative scenic ones. Aldo Leopold's land aesthetic provides a seminal autonomous natural aesthetic theory which may help to awaken our response to the potential of these aesthetically neglected communities.

Leopold shows us that an autonomous natural aesthetic can

involve so much more than the visual appeal of natural environments. A person is in the landscape, i.e., in the natural environment, as the mobile center of a three-dimensional, multi-sensuous experiential continuum. The appreciation of an environment's natural beauty can involve the ears (the sounds of rain, insects, birds, or silence itself), the surface of the skin (the warmth of the sun, the chill of the wind, the texture of grass, rock, sand), the nose and tongue (the fragrance of flowers, the odor of decay, the taste of saps and waters)—as well as the eyes. Most of all it can involve the mind, the faculty of cognition.

Leopold was, apparently, well aware of the primacy of artifactual aesthetics in Western civilization and thus, as an expository device, he approaches natural aesthetics via analogy with the more familiar branch.

He remarks, "our ability to perceive quality in nature begins, as in art, with the pretty [exemplified by, for example, "landscaped" English gardens]. It expands through successive stages of the beautiful [exemplified by, say, the naturally "picturesque" Yosemite Valley, high Alpine "scenery," "sublime" sequoia groves] to values as yet uncaptured by language" (96). Leopold then goes on to capture, in his own compact, descriptive prose, the subtler gamut of aesthetic quality in a nonlandscape which the conventional picturesque natural aesthetic finds plain, if not odious—a crane marsh.

The analogy he draws to art appreciation, i.e., artifactual as opposed to natural aesthetic sensibility, is apt. Among gallery-goers there are also those whose taste is limited to the pretty (to naive, realistic, still-life, and portraiture, for example). Then, there are those capable of appreciating successive stages of the beautiful present in "modern art" (Cezanne, Picasso, or Pollack), whether pretty or not. As in artifactual aesthetics, the capacity to actualize the aesthetic potentialities of land which go beyond the pretty and the picturesque requires some cultivation of sensibility. One must acquire "a refined taste in natural objects":

> The taste for country displays the same diversity in esthetic competence among individuals as the taste for opera, or oils. There are

those who are willing to be herded in droves through "scenic" places; who find mountains grand if they be proper mountains with waterfalls, cliffs, and lakes. To such the Kansas plains are tedious.[7]

For Leopold the Kansas plains are aesthetically exciting less for what is directly seen (or, indeed, otherwise sensuously experienced) than for what is known of their history and biology: Those who have not learned to "read" the land "see the endless corn, but not the heave and grunt of ox teams breaking the prairie. . . . They look at the low horizon, but they cannot see it, as de Vaca did, under the bellies of the buffalo."[8]

In "Marshland Elegy" Leopold beautifully illustrates the impact of an evolutionary biological literacy on perception. Wisconsin's first settlers called sandhill cranes "red shitepokes" for the rusty clay stain their "battleship-gray" feathers acquire in summer (99). The Wisconsin homesteaders saw red shitepokes as just large birds in the way of farm progress. But evolutionary literacy can alter and deepen perception:

> Our appreciation of the crane grows with the slow unraveling of earthly history. His tribe, we now know, stems out of the remote Eocene. The other members of the fauna in which he originated are long since entombed within the hills. When we hear his call we hear no mere bird. We hear the trumpet in the orchestra of evolution. He is the symbol of our untamable past, of that incredible sweep of millennia which underlies the daily affairs of birds and men. (96)

Ecology, as Leopold pictures it, is the biological science which runs at right angles to evolution.[9] Evolution lends to perception a certain depth, "that incredible sweep of millennia," while ecology provides it breadth: Wild things do not exist in isolation from one another. They are "interlocked in one humming community of co-operations and competitions, one biota."[10] Hence the crane, no mere bird, lends "a paleontological patent of nobility" to its marshy habitat (97). We cannot love cranes and hate marshes. The marsh itself is now transformed by the presence of cranes from a "waste," "God-forsaken" mosquito swamp, into a thing of precious beauty.[11]

The crane is also a species native to its marshy habitat. Many pretty plants and animals are not. From the point of view of the land aesthetic, the attractive purple flower of centaurea or the vivid orange of hawkweed might actually spoil rather than enhance a field of (otherwise) native grasses and forbs. Leopold writes lovingly of draba, pasque-flowers, silphium, and many other pretty and not-so-pretty native plants, but with undisguised contempt for peonies, downy chess or cheat grass, foxtail, and other European cultigens and stowaways. He takes delight in the sky dance of the native woodcock and the flight plan of a fleeing partridge, but not (hunter though he may have been) in, say, the gracefully trailing tail-feathers of the imported oriental ring-necked pheasant on the wing.

Ecology, history, paleontology, geology, biogeography—each a form of knowledge or cognition—penetrate the surface provided by direct sensory experience and supply substance to "scenery." Leopold was quite consciously aware of the profound revolution in general sensibility which he was calling for. He contemptuously disparaged "that under-aged brand of esthetics which limits the definition of 'scenery' to lakes and pine trees" (191). "In country a plain exterior often conceals hidden riches." [12] To get at these hidden riches takes more than a gaze at a scenic view through a car window or camera viewfinder. To promote appreciation of nature is "a job not of building roads into lovely country, but of building receptivity into the still unlovely human mind" (176–77).

Leopold's land aesthetic, like his land ethic, is self-consciously informed by evolutionary and ecological biology. It involves a subtle interplay between conceptual schemata and sensuous experience. Experience, as the British Empiricists insisted, informs thought. That is true and obvious to everyone. What is not so immediately apparent is that thought equally and reciprocally informs experience. The "world," as we drink it in through our senses, is first filtered, structured, and arranged by the conceptual framework or cognitive set we bring to it, prior, not necessarily to all experience, but to any *articulate* experience.

This was Kant's great and lasting contribution to philosophy, his self-styled "Copernican revolution" of philosophy. Kant believed that the cognitive conditions of experience were a priori—universal and necessary—but that proved to be a narrowly parochial judgment. The discovery by anthropologists of very different "cultural worlds" or "world views" and subsequent revolutionary changes in Western science affecting even the fundamental experiential parameters of space and time relativized Kant's transcendental "aesthetic" and "logic." His basic revolutionary idea, though, remains very much intact. What one experiences is as much a product of how one thinks as it is the condition of one's senses and the specific content of one's environment.

Leopold is quite consciously aware of the interplay between the creative or active cognitive component of experience and the receptive or passive sensory component. He imagines what Daniel Boone's experience of nature must have been like as an outdoorsman living before the development of an evolutionary-ecological biology:

> Daniel Boone's reaction depended not only on the quality of what he saw, but on the quality of the mental eye with which he saw it. Ecological science has wrought a change in the mental eye. It has disclosed origins and functions for what to Boone were only facts. It has disclosed mechanisms for what to Boone were only attributes. We have no yardstick to measure this change, but we may safely say that, as compared with the competent ecologist of the present day, Boone saw only the surface of things. The incredible intricacies of the plant and animal community—the intrinsic beauty of the organism called America, then in the full bloom of her maidenhood—were as invisible and incomprehensible to Daniel Boone as they are today to Mr. Babbitt. The only true development in American recreational resources is the development of the perceptive faculty in Americans. All of the other acts we grace by that name are, at best, attempts to retard or mask the process of dilution. (173–74).

Thus, while an autonomous natural aesthetic, as I earlier pointed out, must free itself from the prevailing visual bias and involve all sensory modalities, it is not enough simply to open the senses to natural stimuli and enjoy. A complete natural aesthetic, like a complete artifactual aesthetic, shapes and directs sensation, often in surprising ways. It is possible, in certain theoretical contexts, to enjoy and appreciate dissonance in music or the clash of color and distortion of eidetic form in painting. Similarly, in natural

The Land Aesthetic

aesthetics, it is possible to appreciate and relish certain environmental experiences which are not literally pleasurable or sensuously delightful.

For example, a certain northern bog with which I am acquainted is distinguished from the others in its vicinity by the presence of pitcher plants, an endangered species of floral insectivore. I visit this bog at least once each season. The plants themselves are not, by garden standards, beautiful. They are dark red in color, less brilliant than maple leaves in autumn, and they humbly hug the low bog floor. They lie on a bed of sphagnum moss in the deep shade of fifty-foot, ruler-straight tamaracks. To reach the bog I must wade across its mucky moat, penetrate a dense thicket of tag alders, and in summer fight off mosquitoes and black- and deer-flies. My shoes and trousers get wet; my skin gets scratched and bitten. The experience is certainly not spectacular or, for that matter, particularly pleasant; but it is always somehow satisfying aesthetically. The moss bed on which the pitcher plants grow is actually floating. It undulates sensually as I walk through. I smell the sweet decay aroma of the peat and hear the whining insects (in season). I run my finger down and then up the vulva-shaped interior of a pitcher plant's leaf—turned insect trap—to feel the grain of the fibers which keep the insects from crawling out again. It is silky smooth on the way in, bristled on the way out. I look through the trees, beyond, to the adjoining pond. I dig my hand into the moss and bring up a brown rotting mass. I sometimes see a blue heron lift itself off the shore into the air with a single silent stroke of its great wings or see the glint and splash of a northern pike out on the pond. I feel the living bark of the tamaracks, precariously anchored on a floating island. In spring and summer everything is drably green or brown except the sky, and the pitcher plants. Fall is the most colorful, the tamaracks are a "smokey gold." In winter everything shades from black to white. Yet there is a rare music in this place. It is orchestrated and deeply moving.

The beauty of this bog is not serial—an aggregate of interesting objects, like specimens displayed in cases in a natural history museum—nor is it phenomenological—a variety of sensory stimuli or "sense data"; rather, its beauty is a function of the palpable organization and closure of the interconnected living components. The sphagnum moss and the chemical regime it imposes constitute the basis of this small, tight community. The tamaracks are a second major factor. The flora and fauna of the stories between are characteristic of this sort of community, and some, like the pitcher plants, are unique to it. There is a sensible fittingness, a unity there, not unlike that of a symphony or a tragedy. But these connections and relations are not directly sensed in the aesthetic moment, they are *known* and *projected,* in this case by me. It is this conceptual act that completes the sensory experience and causes it to be distinctly aesthetic . . . instead of merely uncomfortable.

Given the Western heritage, it is, perhaps, impossible to express and analyze natural aesthetic experience except by analogy with artifactual aesthetic experience. Leopold's evolutionary-ecological aesthetic is, yielding to this expository necessity, perhaps more akin to aural aesthetics than to visual aesthetics. Few authors have expressed the sense of the familiar metaphor "harmony of nature" with more authority and grace:

> The song of a river ordinarily means the tune that waters play on rock, root, and rapid. . . .
> This song of the waters is audible to every ear, but there is other music in these hills, by no means audible to all. To hear even a few notes of it you must first live here for a long time, and you must know the speech of hills and rivers. Then on a still night, when the campfire is low and the Pleiades have climbed over rimrocks, sit quietly and listen for a wolf to howl, and think hard of everything you have seen and tried to understand. Then you may hear it—a vast pulsing harmony—its score inscribed on a thousand hills, its notes the lives and deaths of plants and animals, its rhythms spanning the seconds and the centuries. (149)

Leopold, in addition to this general Kantian emphasis on the cognitive dimension of natural aesthetic experience, has formulated a quite specialized and somewhat technical natural aesthetic category, the *noumenon,* also ultimately inspired by the philosophy of Kant. To an academic historian of philosophy, Leopold

may seem simply to have misappropriated Kant's term. By *noumenon* Kant meant a purely intelligible object, a thing-in-itself
(*Ding an sich*) which was beyond human ken. Only phenomena
are present to human consciousness, according to Kant. In Leopold's general sense of the term, however, the noumena of land
are quite actual or physical (and therefore, strictly speaking, phenomenal). Nonetheless, in a metaphorical way, they constitute
the "essence" of the countryside. In this sense Leopold's usage
observes the spirit of Kant's definition, if not the letter. Here is
how Leopold introduces the term *noumenon* in *Sand County:*

> The physics of beauty is one department of natural science still in
> the Dark Ages. Not even the manipulators of bent space have tried to
> solve its equations. Everybody knows, for example, that the autumn
> landscape [as we see, even Leopold is not altogether free of the com-
> mon bias] in the north woods is the land, plus a red maple, plus
> a ruffed grouse. In terms of conventional physics, the grouse rep-
> resents only a millionth of either the mass or the energy of an acre.
> Yet subtract the grouse and the whole thing is dead. An enormous
> amount of some kind of motive power has been lost.
>
> It is easy to say that the loss is all in our mind's eye, but is there any
> sober ecologist who will agree? He knows full well that there has
> been an ecological death, the significance of which is inexpressible in
> terms of contemporary science. A philosopher has called this impon-
> derable essence the *numenon* of material things. It stands in contra-
> distinction to *phenomenon,* which is ponderable and predictable, even
> to the tossings and turnings of the remotest star.
>
> The grouse is the numenon of the north woods, the blue jay of the
> hickory groves, the whisky-jack of the muskegs, the piñonero of the
> juniper foothills. . . . (137–38)

And we could go on: the cutthroat trout of high mountain
streams, the sandhill crane of northern marshes, the pronghorn
antelope of the high plains, the loon of glacial lakes, the alligator
of southeastern swamps . . . We might call these noumena more
precisely, though less arrestingly, "aesthetic indicator species."
They supply the hallmark, the imprimatur, to their respective
ecological communities. If they be missing, then the rosy glow
of perfect health, as well as aesthetic excitement, is absent from

the countryside. Like the elusive mountain lion and timber wolf, they need not be seen or heard to grace and enliven their respective habitats. It is enough merely to *know* that they are present.

To sum up, the land aesthetic, desultorily and intermittently developed in *A Sand County Almanac* (and in *Round River*), is a new natural aesthetic, the first, to my knowledge, to be informed by ecological and evolutionary natural history and thus, perhaps, the only genuinely autonomous natural aesthetic in Western philosophical literature: It does not treat natural beauty as subordinate to or derivative from artifactual beauty. However, because natural beauty has traditionally and historically been treated as a reflection of artifactual beauty, the land aesthetic is perforce developed by analogy with artifactual aesthetics. Though more nearly analogous to an aesthetic of music, the land aesthetic is no more aurally biased than visually biased. It involves all sensory modalities equally and indiscriminately.

The popularly prevailing natural aesthetic, the scenic or picturesque aesthetic, frames nature, as it were, and deposits it in "galleries"—the national parks—for most ordinary folk, far from home. We herd in droves to Yellowstone, Yosemite, and the Smokies to gaze at natural beauty and, home again, despise the river bottoms, fallow fields, bogs, and ponds on the back forty. The land aesthetic enables us to mine the hidden riches of the ordinary; it ennobles the commonplace; it brings natural beauty literally home from the hills.

The land aesthetic is sophisticated and cognitive, not naive and hedonic; it delineates a refined taste in natural environments and a cultivated natural sensibility. The basis of such refinement or cultivation is natural history, and more especially, evolutionary and ecological biology. The crane, for example, is no mere bird because of its known, not directly sensed, phylogenetic antiquity; thus the experience of cranes brings an especial aesthetic satisfaction only to those who have a paleontological dimension to their outlook. The experience of a marsh or bog is aesthetically satisfying less for what is literally sensed than for what is known or schematically imagined of its ecology. Leopold enters a qualification, however, to the cognitive stress of the land aesthetic:

The Land Aesthetic

Let no man jump to the conclusion that Babbitt must take his Ph.D. in ecology before he can "see" his country. On the contrary, the Ph.D. may become as callous as an undertaker to the mysteries at which he officiates. Like all real treasures of the mind, perception can be split into infinitely small fractions without losing its quality. The weeds in a city lot convey the same lesson as the redwoods; the farmer may see in his cow-pasture what may not be vouchsafed to the scientist adventuring in the South Seas. Perception, in short, cannot be purchased with either learned degrees or dollars; it grows at home as well as abroad, and he who has a little may use it to as good advantage as he who has much. As a search for perception, the recreational stampede is footless and unnecessary. (174)

Finally and, practically speaking, more important, the land aesthetic is not biased in favor of some natural communities or some places over others. Leopold in his discussion and I in mine have dwelt on wetlands (marshes and bogs) because they are characteristic of Wisconsin (where I also live) and also because—since they are so thoroughly unaesthetic, as measured by conventional canons of scenery—they highlight certain contrasts between the picturesque aesthetic and the land aesthetic. But scenically beautiful environments—alpine communities, for example—are, for that reason, no less land-aesthetically interesting. All biocenoses from arctic tundra to tropical rainforest and from deserts to swamps can be aesthetically appealing upon the land aesthetic. Hence, no matter where one may live, one's environment holds the potential for natural aesthetic experience.

The land aesthetic, thus, complements the land ethic in conservation value theory. The land aesthetic and the land ethic are both equally implied by evolutionary and ecological theory.[13] Together they represent a coherent environmental axiology.

Notes

1. This statement is inaccurate. The academic philosophical community, on the whole, has neither acknowledged nor appreciated it. For a discussion, see J. Baird Callicott, "The Conceptual Foundations of the Land Ethic," in this volume.

2. Roderick Nash in "Aldo Leopold's Intellectual Heritage," in this

volume, does not agree. For a discussion of the limitation of the land ethic to *terrestrial* nature see J. Baird Callicott, "Moral Considerability and Extraterrestrial Life," in *Beyond Space Ship Earth,* ed. Eugene C. Hargrove (San Francisco: Sierra Club Books, 1987).

3. Aldo Leopold, *Round River* (New York: Oxford University Press, 1953), 149.

4. Some recent exceptions include Marjorie Hope Nicholson, *Mountain Gloom, Mountain Glory: The Development of the Aesthetics of the Infinite* (Ithaca: Cornell University Press, 1969); Ian McHarg, *Design with Nature* (Garden City, N.Y.: Doubleday and Co., 1971); Paul Shepard, *Man in the Landscape: A Historic View of the Esthetics of Nature* (New York: Alfred Knopf, 1967); Yi-Fu Tuan, *Topophilia: A Study of Environmental Perception, Attitudes, and Values* (Englewood Cliffs, N.J.: Prentice-Hall, 1974); Mark Sagoff, "On Preserving the Natural Environment," *Yale Law Journal* 84 (1974): 245–67; Ronald Rees, "The Taste for Mountain Scenery," *History Today* 25 (1975): 305–12; Eugene C. Hargrove, "The Historical Foundations of American Environmental Attitudes," *Environmental Ethics* 1 (1979): 209–40; Eugene C. Hargrove, "Anglo-American Land Use Attitudes," *Environmental Ethics* 2 (1980): 121–48; Allen Carlson, "Appreciation and the Natural Environment," *Journal of Aesthetics and Art Criticism* 37 (1979): 267–75; Allen Carlson, "Nature, Aesthetic Judgment, and Objectivity," *Journal of Aesthetics and Art Criticism* 40 (1981): 15–27; Allen Carlson, "Nature and Positive Aesthetics," *Environmental Ethics* 6 (1984); 5–34; Barbara Novak, *Nature and Culture: American Landscape Painting 1825–1875* (New York: Oxford University Press, 1980); J. Baird Callicott, "Aldo Leopold's Land Aesthetic and Agrarian Land Use Values," *Journal of Soil and Water Conservation* 38 (1983): 329–32; Philip G. Terrie, *Forever Wild: Environmental Aesthetics and the Adirondack Forest Preserve* (Philadelphia: Temple University Press, 1985); Holmes Rolston, III, several essays in *Philosophy Gone Wild* (Buffalo, N.Y.: Prometheus Books, 1986).

5. Christopher Hussey, *The Picturesque: Studies in a Point of View* (London: G. P. Putnam's Sons, 1927), 1–2.

6. Rees, in "Mountain Scenery," argues that, if anything, taste in natural "landscapes" has degenerated to what was known in the picturesque aesthetic literature as "a prospect," i.e., a viewpoint or scene.

7. Aldo Leopold, *Round River,* 32–33.

8. Ibid., 33.

9. Ibid., 159.

10. Ibid., 148.

11. Wetlands to those farmers wearing the blinders of the marketplace

The Land Aesthetic

(by no means all farmers) are, in my neck of the woods, regularly referred to as "waste" lands, because they are not in "production," i.e., not cultivated.

12. Aldo Leopold, *Round River,* 33.

13. In case my suggestion that evolutionary and ecological theory has value *implications* may appear to commit the Naturalistic Fallacy, the illicit derivation of value from fact, see J. Baird Callicott, "Hume's Is/Ought Dichotomy and the Relation of Ecology to Leopold's Land Ethic," *Environmental Ethics* 4 (1982): 163–74, for reassurance that it does not.

8 ✴ Building "The Land Ethic"

CURT MEINE

When Aldo Leopold conceived his essay "The Land Ethic" he could not have foreseen the breadth of its influence. A generation has passed since the full essay first appeared in the pages of *A Sand County Almanac,* yet it continues to grow in importance. Students read it. Journalists quote it. Environmentalists live by it. Supreme Court Justices cite it. Scholars criticize it. Many readers have gained their first exposure to an ecological world view through it. Recently it has become a standard starting point for scholarly discussions of environmental ethics. In short, "The Land Ethic" has helped lead a generation in reassessing its relationship to the natural environment.

"The Land Ethic" is also Aldo Leopold's most enduring expression of his personal convictions. Written at the close of a varied career, the essay is a synthesis of his ideas and experience. Leopold was, in some respects, a paradoxical figure: a scientist who wrote poetry, a scholar who was most comfortable in the field, a conservative man who came to advocate a revolutionary idea. In his own life he was able to maintain the connection between emotion and reason; "The Land Ethic" was the final proof of that connection.

This essay is a narrative abstract of the author's Master of Science thesis (submitted and accepted at the University of Wisconsin–Madison in 1983), "Building the Land Ethic: A History of Aldo Leopold's Most Important Essay."

Building "The Land Ethic"

In light of these considerations, there is need for a more critical understanding of the process by which Leopold put together the words, thoughts, and arguments of "The Land Ethic."

"The Land Ethic" was first published in Part III of *A Sand County Almanac,* "The Upshot." It is the climactic essay of this third section—in effect, the upshot of "The Upshot." It is a summary piece in which the issues raised explicitly and implicitly throughout the book are finally distilled.[1]

As published, "The Land Ethic" reads as a single coherent statement which the author might have written over a short period of time. In fact, the essay was written in four phases over a fourteen-year period.

The opening parts of "The Land Ethic" are also the earliest written. Leopold derived most of the essay's first two subsections from an address he gave before a meeting of the American Association for the Advancement of Science in Las Cruces, New Mexico, on May 1, 1933. In this address, entitled "The Conservation Ethic," Leopold discussed the inability of economics-as-usual to deal with the basic problems of land abuse. The predominant economic "isms," Leopold wrote, lacked "any vital proposal for adjusting men and machines to land."[2] The corollary to this, he ventured, was that conservation contained seeds from which such a proposal might emerge. This address was first published in the *Journal of Forestry* in October of the same year.

A large portion of "The Land Ethic" draws on another address by Leopold, "A Biotic View of Land," delivered before a joint meeting of the Ecological Society of America and the Society of American Foresters in Milwaukee on June 21, 1939. It has been called "Leopold's earliest comprehensive statement of the new ecological viewpoint," and reflects Leopold's immersion in ecological research and reasoning during the 1930s.[3] In it, Leopold explained the ecological dynamics of the land community and argued for an approach to land use that respected these dynamics. He employed the land pyramid model to illustrate the "smooth functioning" of the community, asked whether human alterations of that community could be less "violent," and finally pointed out trends in conservation that suggested that such "violence" in fact could be reduced. "A Biotic View of Land" first appeared in the *Journal of Forestry* in September 1939.

Leopold's discussion of the ecological conscience is taken from a third address, of that name, which was given before the Garden Club of America on June 27, 1947, in Minneapolis. In this speech, Leopold forcefully stated that, ultimately, conservation must spring from a sense of individual responsibility for the general health of the land. Citing four cases from his own experience in Wisconsin, Leopold held that conservation by government agency can only do so much, and that sound land use must rest upon the land user's own "ecological conscience." This address was first published in the *Bulletin of the Garden Club of America* in September 1947.

About one half of "The Land Ethic" was newly written specifically for the essay. A rough draft of "The Ecological Conscience" bears the date of that address's delivery, "6-27-47." The first reference to an essay called "The Land Ethic" appears in the "Foreword [to *Great Possessions*)" dated "7-31-47."[4] These texts allow us to point to July 1947 as the period in which Leopold, then in the final stages of work on the book as a whole, finally assembled and synthesized "The Land Ethic." The new portions of the essay were undoubtedly composed at this time.

"The Land Ethic," then, is neither a wholly new essay, nor a mere compilation of older material. Rather, it is a new expression of older thoughts, some of which were altered in the process, some not. In changing and bringing together previous statements, and in adding fresh ideas, Leopold created an argument stronger than the sum of its original parts, and in this, perhaps, lies the essay's enduring quality.

A careful comparison of "The Land Ethic" with its three antecedents reveals three categories of revisions. First, there are purely stylistic revisions. These occur throughout the essay and range from the grammatical and cosmetic to the rhetorical and poetic. Second, there are changes of emphasis and clarification. These also occur throughout the essay, but are especially prominent in the subsections The Community Concept, The Land Pyramid, and Land Health and the A-B Cleavage. Third, there are changes that indicate deliberate shifts in tone and thought. The Land Pyramid is significantly altered in this regard, as, to a lesser degree, is The Community Concept.

In sum, these changes reveal Leopold speaking with a more confident voice, aiming for accuracy of expression, pulling his punches at several key points, and allowing his argument to speak for itself. There is an evident effort to broaden, in time and space, the scope of his ideas. Leopold was more strident in each of the original addresses; he is more direct in the final essay. The new portions of the essay share this measured quality.

"The Conservation Ethic" dealt primarily with the cultural and historical aspects of conservation; "A Biotic View of Land" dealt with the biological aspects, and "The Ecological Conscience" with what might be called the "personal" aspects. In each of these spheres, Leopold experienced a positive shift over the last fifteen years of his life. "The Conservation Ethic" was almost entirely a criticism of economic standards for determining land use; in "The Land Ethic" Leopold more actively asserts ethical standards. Leopold's preoccupation in the 1930s with what he termed "land sickness" and "violent use of land" gave way to a conception of "land health" and a call for "gentler and more objective" criteria in our use of land. Finally, Leopold's criticism of the role of government during the New Deal became its own inverse: a declaration of the individual's responsibilities in living on the land. By fusing these themes together, or, perhaps more accurately, by recognizing that they could not be severed, Leopold built his land ethic.

While "The Conservation Ethic" stands as Leopold's first important publication on the philosophical dimensions of conservation, only a small portion of the original actually survived Leopold's later incorporation of it into "The Land Ethic." His purpose in the original essay was to place the conservation movement firmly at the forefront of a historical trend—the tendency of ethical considerations to supercede purely economic ones. In conservation's tools and ideas Leopold saw the potential for a new, deeper, and more enduring relationship with land. To firm up this thesis, Leopold had first to explain in greater detail how ethical systems expanded over time, and how conservation fit into the process. By 1947 Leopold had strengthened his supporting arguments, so that all he retained from "The Conservation Ethic" was the original introductory explanation of this "extension of ethics."

Leopold opens "The Land Ethic" with a three-paragraph allusion to Homer's *Odyssey,* illustrating his point that mankind's ethical structure can and does evolve. Taken directly from the opening of "The Conservation Ethic," this passage not only displays Leopold's familiarity with classical literature but emphasizes his historical perspective—already firmly established in the previous sections of *A Sand County Almanac*—and announces his intention of making a statement about cultural evolution.

The first of the seven subtitled sections of "The Land Ethic" is The Ethical Sequence. In a concise six paragraphs, Leopold addresses a complex subject: the relationship between ethics and biology and the course of that relationship through history. All of the text of The Ethical Sequence section is taken from "The Conservation Ethic." In the original it is not given a separate subtitle, but follows directly after the opening allusion to the *Odyssey.*

Several of the revisions of this text which Leopold made in 1947 are worth pointing out. Leopold changed the first paragraph of what became The Ethical Sequence only slightly, but significantly. "Biological," "biologically," and "biologist" in the original become "ecological," "ecologically," and "ecologist" in the reworked version. These substitutions seem to reflect Leopold's greater willingness to employ the term "ecology"—a word not in general circulation in 1933—as well as Leopold's own development as an ecologist in the intervening years.[5]

Leopold opened the original version of the third paragraph of The Ethical Sequence with two reiterative sentences on the evolution of ethics. They are deleted from "The Land Ethic." The only other substantial change in this paragraph is the substitution of "The Golden Rule" for the original "Christianity," reflecting, if nothing else, an attempt to be more specific.

The first sentence of the fifth paragraph is reworded, and two sentences describing specific instances of social injustice in China,

Germany, and Greece are deleted. His explicit claim in the 1933 version was that ecological "injustices" might arouse as much moral outrage in us as injustices in human society. This is a strong statement even today, and no less difficult to defend. Leopold may have thought that he was overstating his case at this point and digressing from the main line of the argument.

The original conclusion of The Ethical Sequence contained several key points about the relationship between science and ethics. "Some scientists," Leopold wrote in 1933, "will dismiss this matter [of a conservation ethic] forthwith, on the ground that ecology has no relation to right and wrong. To such I reply that science, if not philosophy, should by now have made us cautious about dismissals. . . . No ecologist can deny that our land relation involves penalties and rewards which the individual does not see, and needs modes of guidance which do not yet exist. Call these what you will, science cannot escape its part in forming them." He deleted these statements in 1947, an indication that he was keenly aware that he was entering dangerous philosophical waters. He knew well that, in the orthodox views of both philosophy and science, ethics and scientific inquiry were not even on speaking terms with one another. Leopold's firm belief was always that science and ethics, though distinct, could not be separated. In a sense, that was the entire point of "The Land Ethic."[6]

On the whole, Leopold's revisions of The Ethical Sequence were not extensive. He trimmed it a bit, removed some passages and clarified others, making it a more appropriate introduction to an essay wider in scope than the original. The changes suggest that Leopold was still trying to clarify his ideas on the natural basis of ethics, reflecting a broader ecological view of that basis. The several deletions in this section keep the discussion focused on the fundamental relationship among ethics, evolution, and ecology. The general purpose and plan of this section remain the same as in the original: to inform the reader that there is a historical growth of ethics toward an expanded conception of the social environment. Leopold explores that expanded community in detail in the next section, The Community Concept.

This second subtitled section of "The Land Ethic" defines and illustrates the land community. The first half of The Community Concept was newly written in 1947, and it contains two of the essay's more often-quoted passages:

> The land ethic simply enlarges the boundaries of the community to include soils, waters, plants, and animals, or collectively: the land. (204)

> In short, a land ethic changes the role of *Homo sapiens* from conqueror of the land-community to plain member and citizen of it. It implies respect for his fellow-members, and also respect for the community as such. (204)

To substantiate this view of land as a community and man as a too-often-unaware member of it, Leopold suggests that we look upon history not as a story of human enterprise alone, but as a complex tale of interaction between humanity's ambition and technique and the land's natural diversity and dynamism. The second half of the section, borrowed again from "The Conservation Ethic" of 1933, gives examples of this approach to history, citing European settlement of two very different North American environments: the mesic "canelands" of Kentucky and the arid ranges of the Southwest.

In 1933, Leopold had followed his discussion of the extension of ethics not with a broadened definition of the community per se, but with a section specifically subtitled Ecology—Its Role in History. When he revised the text in the summer of 1947, he first wrote out his new definition of community and only then returned to "The Conservation Ethic" for historical examples.

The original draft of the first eight paragraphs of The Community Concept is not extant, but it is safe to assume that this more abstract, philosophical passage was written just as Leopold was piecing the essay together in 1947. There are four such untraceable passages in "The Land Ethic," and all of them contain at least one reference to "a (or the) land ethic"; in the remainder of the essay, all of which *can* be traced to the three previous articles, the phrase does not occur even once. In fact, Leopold had never before used the term in his published works.

The assembling of The Community Concept exemplifies the

process Leopold went through in putting together "The Land Ethic" as a whole. The half of the section that had been on paper for fourteen years was still relevant, but the theme of that passage—a conservationist's view of human history—no longer stood on its own. In the revision it supports a more general idea which in the intervening years had become more important to Leopold—an ecologist's view of the land community—and which is emphasized in a brand new passage. The editorial changes Leopold made in reworking "The Conservation Ethic" for this section are mostly minor, but they reveal basic changes in his approach: he softened his words and made a definite effort to apply his argument to additional times and places.

Having defined his concept of the land community, Leopold in his next section, The Ecological Conscience, suggests that as long as conservation is construed to entail merely the wise use of natural resources, it will remain limited in overall effectiveness. Conservation that emphasizes only external changes, legislative acts, and technological fixes, Leopold argues, neglects the other half of the equation: the attitudes and values of the individual. He cites, as an example, the problems the Soil Conservation Service faced in the 1930s when it tried to encourage the farmers of southwestern Wisconsin to adopt soil conservation techniques. Leopold concluded that the conservation movement had not yet challenged individual values and beliefs to any significant degree.

This was a standard theme in Leopold's writing, particularly after the advent of the New Deal. He finally summarized his views in his address "The Ecological Conscience" on June 27, 1947. In the weeks that followed, he condensed the speech and included it in "The Land Ethic." Although Leopold borrowed the title "The Ecological Conscience" for use in "The Land Ethic," he does not there give a clear definition of the term. In the original, he had: "Ecology is the science of communities," he wrote, "and the ecological conscience is therefore the ethics of community life." [7]

Leopold originally presented "The Ecological Conscience" as a speech illustrated with slides. The bulk of the address involved four case histories of ineffective conservation efforts (one of which was the soil conservation example). In condensing the original address, Leopold's editorial decisions were primarily rhetorical. The speech had been one of the most forcefully worded of his career, and he made an evident effort to tone it down, to move from stridency to calm concern. This shift is exemplified in the opening sentence of the section. He had originally written that "Everyone ought to be dissatisfied with the slow spread of conservation to the land." He replaced this somewhat sanctimonious assertion with his classic statement, "Conservation is a state of harmony between men and land (207)."

The enduring contribution of "The Land Ethic" to environmental philosophy is its clear conviction that the free individual must be responsible for and responsive to the land he or she lives upon. While Leopold had addressed this theme before, only with "The Ecological Conscience" did he reduce that conviction to its essentials. In "The Land Ethic," compiled soon after, we find it in a broader context. The presence of the surrounding argument and the desire to persuade a larger, less committed public may account for the "toning down."

"The Ecological Conscience," especially in its full-length address form, challenges the individual to assume responsibility for the health and decent use of the natural community. In Substitutes for a Land Ethic, the next section, Leopold describes what happens when economics (or "expediency") alone determines how that responsibility shall be discharged: most elements of the community (Leopold mentions, as examples, songbirds, predators, raptors, and noncommercial tree species) are ignored or even actively eliminated; whole communities (bogs, marshes, dunes) are read out of the purely economic landscape; government must step in to take over "all necessary jobs that private landowners fail to perform," a development Leopold considered cumbersome at best, ineffective at worst (213). According to Leopold, these situations will be remedied only when the private landowner assumes an ethical attitude toward land.

This is the second of the four portions of "The Land Ethic" that apparently were written as the essay was being synthesized. No part of this section shows up in Leopold's previously published works and no rough draft of it exists among Leopold's collected unpublished papers. The telltale phrase "land ethic" appears four times in this section, including once, of course, in the title.

The longest and least quoted section of "The Land Ethic" is The Land Pyramid. Here Leopold explains and applies the ecological model of the biotic pyramid. The first part of this section discusses the structure and function of the naturally evolved community, using basic ecological constructs: trophic levels, food chains, diversity and stability, and energy flow. Leopold next stresses how changes, evolutionary and technological, affect the smooth functioning of the ecosystem. Finally, Leopold gives a brief survey of human impacts on the world's biotas and assesses the effects of those conversions. He concludes that "the less violent the man-made changes, the greater the probability of successful readjustment in the pyramid" (220). Leopold characteristically ends the section by emphasizing our dim awareness of ecological dynamics.

Virtually all of The Land Pyramid is drawn from Leopold's 1939 "A Biotic View of Land." Its condensed version in "The Land Ethic" provides the scientific backbone to Leopold's argument.

Of all the sections of "The Land Ethic" that were based on previously published material, none was so minutely and carefully revised as this one. Most of the additions, deletions, substitutions, and rephrasings seem minor at the outset, but collectively they show Leopold striving for a concise expression of the complex scientific rationale behind "The Land Ethic." The heart of the discussion is Leopold's explanation of the relationship between structure and function in the ecosystem, an issue that remains a central point of debate among ecologists. The changes in this section show a sharper understanding of both stability and diversity, but no definitive conclusion about their relationship is reached. Several of the changes here also indicate that Leopold is speaking in a voice that is at once more confident and less aggressive. We see, again, an attempt to broaden the discussion to global dimensions. Above all, Leopold was aiming to clarify and focus his explanation.

The Land Pyramid explains in basic terms the structure and function of the land community. Land Health and the A-B Cleavage, the section that follows, explores conservation's internal tension as it tries to "understand and preserve" land health, defined by Leopold as the capacity for self-renewal. In three fields—forestry, wildlife management, and agriculture—Leopold notes the same division between "man the conqueror" and "man the biotic citizen": the former sees land solely as a producer of commodities, the latter sees land as a diverse and integrated whole.

This is the third of the four passages that were newly written for the final essay in 1947. No original draft of it has been found. There is clear evidence, however, that it is a reworking of the conclusion to "A Biotic View of Land," although the parallels between the original and the revision are not as obvious as in The Land Pyramid.

The first paragraph of this section is obviously an addition. The terms "land ethic" and "ecological conscience" are used in the first sentence. The second paragraph begins with the statement, "Conservationists are notorious for their dissensions"; the outline Leopold follows in illustrating these "dissensions"—in forestry, wildlife, and agriculture—is the same as in "A Biotic View." The passage on agriculture, in particular, closely parallels that of the 1939 essay and contains several of the same phrases. The passages on wildlife both make mention of exotics, rarities, and wildflowers. The passages on forestry both bemoan utilitarian forestry's predilection for growing trees "like cabbages" (one of Leopold's favorite images). The section concludes with a few lines from one of Leopold's favorite poems, used previously in "The Conservation Ethic."[8]

In its revised form, this section shows a much more focused sense of the divisions within and among the branches of conservation. In "A Biotic View of Land" Leopold was primarily concerned with the ecological concept of land; the varying attitudes of conservationists toward land were of secondary importance at the time. For years, he had wrestled with the fact that conservation was divided "between those who see utility and beauty in the biota as a whole, and those who see utility and beauty only in pheasants or trout." In "The Land Ethic," he decided to draw explicitly the line between them.[9]

The Outlook is the last section of "The Land Ethic." Leopold assesses here the prospects for cultivating an ethical relationship with land, suggests what might be done to promote that relationship, and concludes with a plea for a deeper understanding of land.

This section is the fourth newly written part of the essay. Again, no original drafts have turned up, but we may assume that Leopold would write the conclusion only after assembling the essay. The identifying term "land ethic" is mentioned five times in this section. Leopold reiterates his themes: the need for ecological awareness; the limits of education in providing that awareness; the limits of economics in acting upon it; the land ethic as the next step in an evolutionary sequence; the need for more basic understanding of ecological processes; the role of the ecological conscience. All have been mentioned at previous points in the essay, and all return in this summary.

Only two passages come from previous writings. The final image—". . . remodeling the Alhambra with a steam shovel"—is borrowed from "The Conservation Ethic" and expanded in this concluding paragraph.[10]

The other passage is more critical. *A Sand County Almanac* is the climax, in many ways, of Aldo Leopold's career; "The Land Ethic" is the climax of *A Sand County Almanac;* this paragraph is the climax of "The Land Ethic":

> The "key-log" which must be moved to release the evolutionary process for an ethic is simply this: quit thinking about decent land-use as solely an economic problem. Examine each question in terms of what is ethically and esthetically right, as well as what is economically expedient. A thing is right when it tends to preserve the integrity, stability, and beauty of the biotic community. It is wrong when it tends otherwise. (224–25)

No other passage from "The Land Ethic" has been so often quoted or so carefully scrutinized. It is the focus of the entire essay. It originally appeared at the conclusion of "The Ecological Conscience" in a slightly different form:

The practice of conservation must spring from a conviction of what is ethically and esthetically right, as well as what is economically expedient. A thing is right only when it tends to preserve the integrity, stability, and beauty of the community, and the community includes the soil, waters, fauna, and flora, as well as people.[11]

This, in turn, was based on a slightly different original draft in which the phrases "and esthetically" and "and beauty" were absent. Leopold later penciled these phrases in. While he is definitely erecting ethical standards in this statement, according to the criteria of integrity and stability, it seems he was more concerned with touching a broad base of human perceptions of the environment than with developing a hard-and-fast set of timeless standards. Leopold apparently saw the need to address the aesthetically minded portion of his audience. That his own mind was focused on the fusion of ethical use and beauty during this period is also indicated in the opening line of the "Foreword [to Great Possessions]" of July 31, 1947: "These essays," he wrote, "deal with the ethics and esthetics of land."[12] One suspects that he would have welcomed the subsequent quotation and criticism of this significant passage.

The simplicity of "The Land Ethic" masks the fact that it was actually the final product of a long, careful, and intensely personal process of observation, action, and reflection. Others tried to articulate a comparable philosophy of conservation during Leopold's lifetime, and many have tried since, but "The Land Ethic" remains a notable success. Because of its firm basis in the observed world, it is more accessible than most formal philosophy. Because of its personal tone, it exhibits more understanding than most attempts to derive a philosophy from ecology. Because of its intellectual breadth, it is more comprehensive than most conservation literature. In the time that has passed since the essay first appeared, philosophers, natural scientists, and the lay public have begun to discuss the issues that Leopold saw so clearly in 1947. He remains one of the few who can talk to all with authority.

Building "The Land Ethic"

Notes

1. For further discussion of the structure of *A Sand County Almanac,* see Peter A. Fritzell, "The Conflicts of an Ecological Conscience," and Dennis Ribbens, "The Making of *A Sand County Almanac,*" both in this volume.

2. Aldo Leopold, "The Conservation Ethic," *Journal of Forestry* 31 (October 1933): 639–40.

3. Susan Flader, *Thinking Like a Mountain: Aldo Leopold and the Evolution of an Ecological Attitude toward Deer, Wolves, and Forests* (Columbia: University of Missouri Press, 1974), 30–31.

4. This is found in the Leopold Papers of the University of Wisconsin–Madison Archives, LP 6B16, under the file, "Literary and Philosophical, 1940–1948." It is included in the appendix of this volume.

5. See J. Baird Callicott, "The Conceptual Foundations of the Land Ethic," in this volume. He suggests that these changes were, if anything, ill advised.

6. For a discussion of this problem see J. Baird Callicott, "Hume's Is/Ought Dichotomy and the Relation of Ecology to Leopold's Land Ethic," *Environmental Ethics* 4 (1982): 163–74.

7. Aldo Leopold, "The Ecological Conscience," *Bulletin of the Garden Club of America* (September 1947): 45.

8. This quote, which Leopold used several times in his speeches and articles, is from Edward Arlington Robinson's 1927 long poem *Tristram* (New York: The Macmillan Company, 1927), 83.

9. Aldo Leopold, "A Biotic View of Land," *Journal of Forestry* 37 (1939): 729.

10. Aldo Leopold, "The Conservation Ethic," 637.

11. Aldo Leopold, "The Ecological Conscience," 52.

12. Aldo Leopold, "Foreword [to *Great Possessions*]," in this volume.

9 ★ The Conceptual Foundations of the Land Ethic

J. BAIRD CALLICOTT

> The two great cultural advances of the past century were the Darwinian theory and the development of geology. . . . Just as important, however, as the origin of plants, animals, and soil is the question of how they operate as a community. That task has fallen to the new science of ecology, which is daily uncovering a web of interdependencies so intricate as to amaze—were he here—even Darwin himself, who, of all men, should have least cause to tremble before the veil. (Aldo Leopold, fragment 6B16, no. 36, Leopold Papers, University of Wisconsin–Madison Archives)

As Wallace Stegner observes, *A Sand County Almanac* is considered "almost a holy book in conservation circles," and Aldo Leopold a prophet, "an American Isaiah." And as Curt Meine points out, "The Land Ethic" is the climactic essay of *Sand County,* "the upshot of 'The Upshot.'" [1] One might, therefore, fairly say that the recommendation and justification of moral obligations on the part of people to nature is what the prophetic *A Sand County Almanac* is all about.

But, with few exceptions, "The Land Ethic" has not been favorably received by contemporary academic philosophers. Most have ignored it. Of those who have not, most have been either nonplussed or hostile. Distinguished Australian philosopher John Passmore dismissed it out of hand, in the first book-length academic discussion of the new philosophical subdiscipline called "environmental ethics." [2] In a more recent and more deliberate discussion, the equally distinguished Australian philosopher H. J. McCloskey patronized Aldo Leopold and saddled "The Land Ethic" with various far-fetched "interpretations." He concludes that "there is a real problem in attributing a coherent meaning to Leopold's statements, one that exhibits his land ethic as representing a major advance in ethics rather than a retrogres-

The Conceptual Foundations of the Land Ethic

sion to a morality of a kind held by various primitive peoples."[3] Echoing McCloskey, English philosopher Robin Attfield went out of his way to impugn the philosophical respectability of "The Land Ethic." And Canadian philosopher L. W. Sumner has called it "dangerous nonsense."[4] Among those philosophers more favorably disposed, "The Land Ethic" has usually been simply quoted, as if it were little more than a noble, but naive, moral plea, altogether lacking a supporting theoretical framework—i.e., foundational principles and premises which lead, by compelling argument, to ethical precepts.

The professional neglect, confusion, and (in some cases) contempt for "The Land Ethic" may, in my judgment, be attributed to three things: (1) Leopold's extremely condensed prose style in which an entire conceptual complex may be conveyed in a few sentences, or even in a phrase or two; (2) his departure from the assumptions and paradigms of contemporary philosophical ethics; and (3) the unsettling practical implications to which a land ethic appears to lead. "The Land Ethic," in short, is, from a philosophical point of view, abbreviated, unfamiliar, and radical.

Here I first examine and elaborate the compactly expressed abstract elements of the land ethic and expose the "logic" which binds them into a proper, but revolutionary, moral theory. I then discuss the controversial features of the land ethic and defend them against actual and potential criticism. I hope to show that the land ethic cannot be ignored as merely the groundless emotive exhortations of a moonstruck conservationist or dismissed as entailing wildly untoward practical consequences. It poses, rather, a serious intellectual challenge to business-as-usual moral philosophy.

"The Land Ethic" opens with a charming and poetic evocation of Homer's Greece, the point of which is to suggest that today land is just as routinely and remorsely enslaved as human beings then were. A panoramic glance backward to our most distant cultural origins, Leopold suggests, reveals a slow but steady moral development over three millennia. More of our relationships and activities ("fields of conduct") have fallen under the aegis of moral principles ("ethical criteria") as civilization has grown and matured. If moral growth and development continue, as not only a synoptic review of history, but recent past experience suggest that it will, future generations will censure today's casual and universal environmental bondage as today we censure the casual and universal human bondage of three thousand years ago.

A cynically inclined critic might scoff at Leopold's sanguine portrayal of human history. Slavery survived as an institution in the "civilized" West, more particularly in the morally self-congratulatory United States, until a mere generation before Leopold's own birth. And Western history from imperial Athens and Rome to the Spanish Inquisition and the Third Reich has been a disgraceful series of wars, persecutions, tyrannies, pogroms, and other atrocities.

The history of moral practice, however, is not identical with the history of moral consciousness. Morality is not descriptive; it is prescriptive or normative. In light of this distinction, it is clear that today, despite rising rates of violent crime in the United States and institutional abuses of human rights in Iran, Chile, Ethiopia, Guatemala, South Africa, and many other places, and despite persistent organized social injustice and oppression in still others, moral consciousness is expanding more rapidly now than ever before. Civil rights, human rights, women's liberation, children's liberation, animal liberation, etc., all indicate, as expressions of newly emergent moral ideals, that ethical consciousness (as distinct from practice) has if anything recently accelerated—thus confirming Leopold's historical observation.

Leopold next points out that "this extension of ethics, so far studied only by philosophers"—and therefore, the implication is clear, not very satisfactorily studied—"is actually a process in ecological evolution" (202). What Leopold is saying here, simply, is that we may understand the history of ethics, fancifully alluded to by means of the Odysseus vignette, in biological as well as philosophical terms. From a biological point of view, an ethic is "a limitation on freedom of action in the struggle for existence" (202).

I had this passage in mind when I remarked that Leopold

manages to convey a whole network of ideas in a couple of phrases. The phrase "struggle for existence" unmistakably calls to mind Darwinian evolution as the conceptual context in which a biological account of the origin and development of ethics must ultimately be located. And at once it points up a paradox: Given the unremitting competitive "struggle for existence" how could "limitations on freedom of action" ever have been conserved and spread through a population of *Homo sapiens* or their evolutionary progenitors?

For a biological account of ethics, as Harvard social entomologist Edward O. Wilson has recently written, "the central theoretical problem . . . [is] how can altruism [elaborately articulated as morality or ethics in the human species], which by definition reduces personal fitness, possibly evolve by natural selection?"[5] According to modern sociobiology, the answer lies in kinship. But according to Darwin—who had tackled this problem himself "exclusively from the side of natural history" in *The Descent of Man*—the answer lies in society.[6] And it was Darwin's classical account (and its divers variations), from the side of natural history, which informed Leopold's thinking in the late 1940s.

Let me put the problem in perspective. How, we are asking, did ethics originate and, once in existence, grow in scope and complexity?

The oldest answer in living human memory is theological. God (or the gods) imposes morality on people. And God (or the gods) sanctions it. A most vivid and graphic example of this kind of account occurs in the Bible when Moses goes up on Mount Sinai to receive the Ten Commandments directly from God. That text also clearly illustrates the divine sanctions (plagues, pestilences, droughts, military defeats, etc.) for moral disobedience. Ongoing revelation of the divine will, of course, as handily and as simply explains subsequent moral growth and development.

Western philosophy, on the other hand, is almost unanimous in the opinion that the origin of ethics in human experience has somehow to do with human reason. Reason figures centrally and pivotally in the "social contract theory" of the origin and nature of morals in all its ancient, modern, and contemporary expressions from Protagoras, to Hobbes, to Rawls. Reason is the well-spring of virtue, according to both Plato and Aristotle, and of categorical imperatives, according to Kant. In short, the weight of Western philosophy inclines to the view that we are moral beings because we are rational beings. The ongoing sophistication of reason and the progressive illumination it sheds upon the good and the right explain "the ethical sequence," the historical growth and development of morality, noticed by Leopold.

An evolutionary natural historian, however, cannot be satisfied with either of these general accounts of the origin and development of ethics. The idea that God gave morals to man is ruled out in principle—as any supernatural explanation of a natural phenomenon is ruled out in principle in natural science. And while morality might *in principle* be a function of human reason (as, say, mathematical calculation clearly is), to suppose that it is so *in fact* would be to put the cart before the horse. Reason appears to be a delicate, variable, and recently emerged faculty. It cannot, under any circumstances, be supposed to have evolved in the absence of complex linguistic capabilities which depend, in turn, for their evolution upon a highly developed social matrix. But we cannot have become social beings unless we assumed limitations on freedom of action in the struggle for existence. Hence we must have become ethical before we became rational.

Darwin, probably in consequence of reflections somewhat like these, turned to a minority tradition of modern philosophy for a moral psychology consistent with and useful to a general evolutionary account of ethical phenomena. A century earlier, Scottish philosophers David Hume and Adam Smith had argued that ethics rest upon feelings or "sentiments"—which, to be sure, may be both amplified and informed by reason.[7] And since in the animal kingdom feelings or sentiments are arguably far more common or widespread than reason, they would be a far more likely starting point for an evolutionary account of the origin and growth of ethics.

Darwin's account, to which Leopold unmistakably (if elliptically) alludes in "The Land Ethic," begins with the parental and filial affections common, perhaps, to all mammals.[8] Bonds of affection and sympathy between parents and offspring permitted the formation of small, closely kin social groups, Darwin argued. Should the parental and filial affections bonding family members chance to extend to less closely related individuals, that would permit an enlargement of the family group. And should the newly extended community more successfully defend itself and/ or more efficiently provision itself, the inclusive fitness of its members severally would be increased, Darwin reasoned. Thus, the more diffuse familial affections, which Darwin (echoing Hume and Smith) calls the "social sentiments," would be spread throughout a population.[9]

Morality, properly speaking—i.e., morality as opposed to mere altruistic instinct—requires, in Darwin's terms, "intellectual powers" sufficient to recall the past and imagine the future, "the power of language" sufficient to express "common opinion," and "habituation" to patterns of behavior deemed, by common opinion, to be socially acceptable and beneficial.[10] Even so, ethics proper, in Darwin's account, remains firmly rooted in moral feelings or social sentiments which were—no less than physical faculties, he expressly avers—naturally selected, by the advantages for survival and especially for successful reproduction, afforded by society.[11]

The protosociobiological perspective on ethical phenomena, to which Leopold as a natural historian was heir, leads him to a generalization which is remarkably explicit in his condensed and often merely resonant rendering of Darwin's more deliberate and extended paradigm: Since "the thing [ethics] has its origin in the tendency of interdependent individuals or groups to evolve modes of co-operation, . . . all ethics so far evolved rest upon a single premise: that the individual is a member of a community of interdependent parts" (202–3).

Hence, we may expect to find that the scope and specific content of ethics will reflect both the perceived boundaries and actual structure or organization of a cooperative community or society. *Ethics and society or community are correlative.* This single, simple principle constitutes a powerful tool for the analysis of moral natural history, for the anticipation of future moral development (including, ultimately, the land ethic), and for systematically deriving the specific precepts, the prescriptions and proscriptions, of an emergent and culturally unprecedented ethic like a land or environmental ethic.

Anthropological studies of ethics reveal that in fact the boundaries of the moral community are generally coextensive with the perceived boundaries of society.[12] And the peculiar (and, from the urbane point of view, sometimes inverted) representation of virtue and vice in tribal society—the virtue, for example, of sharing to the point of personal destitution and the vice of privacy and private property—reflects and fosters the life way of tribal peoples.[13] Darwin, in his leisurely, anecdotal discussion, paints a vivid picture of the intensity, peculiarity, and sharp circumscription of "savage" mores: "A savage will risk his life to save that of a member of the same community, but will be wholly indifferent about a stranger."[14] As Darwin portrays them, tribespeople are at once paragons of virtue "within the limits of the same tribe" and enthusiastic thieves, manslaughterers, and torturers without.[15]

For purposes of more effective defense against common enemies, or because of increased population density, or in response to innovations in subsistence methods and technologies, or for some mix of these or other forces, human societies have grown in extent or scope and changed in form or structure. Nations— like the Iroquois nation or the Sioux nation—came into being upon the merger of previously separate and mutually hostile tribes. Animals and plants were domesticated and erstwhile hunter-gatherers became herders and farmers. Permanent habitations were established. Trade, craft, and (later) industry flourished. With each change in society came corresponding and correlative changes in ethics. The moral community expanded to become coextensive with the newly drawn boundaries of societies and the representation of virtue and vice, right and wrong, good and evil, changed to accommodate, foster, and preserve the economic and institutional organization of emergent social orders.

Today we are witnessing the painful birth of a human supercommunity, global in scope. Modern transportation and communication technologies, international economic interdependencies, international economic entities, and nuclear arms have brought into being a "global village." It has not yet become fully formed and it is at tension—a very dangerous tension—with its predecessor, the nation-state. Its eventual institutional structure, a global federalism or whatever it may turn out to be, is, at this point, completely unpredictable. Interestingly, however, a corresponding global human ethic—the "human rights" ethic, as it is popularly called—has been more definitely articulated.

Most educated people today pay lip service at least to the ethical precept that all members of the human species, regardless of race, creed, or national origin, are endowed with certain fundamental rights which it is wrong not to respect. According to the evolutionary scenario set out by Darwin, the contemporary moral ideal of human rights is a response to a perception—however vague and indefinite—that mankind worldwide is united into one society, one community—however indeterminate or yet institutionally unorganized. As Darwin presciently wrote:

> As man advances in civilization, and small tribes are united into larger communities, the simplest reason would tell each individual that he ought to extend his social instincts and sympathies to all the members of the same nation, though personally unknown to him. This point being once reached, there is only an artificial barrier to prevent his sympathies extending to the men of all nations and races. If, indeed, such men are separated from him by great differences of appearance or habits, experience unfortunately shows us how long it is, before we look at them as our fellow-creatures.[16]

According to Leopold, the next step in this sequence beyond the still incomplete ethic of universal humanity, a step that is clearly discernible on the horizon, is the land ethic. The "community concept" has, so far, propelled the development of ethics from the savage clan to the family of man. "The land ethic simply enlarges the boundary of the community to include soils, waters, plants, and animals, or collectively: the land" (204).

As the foreword to *Sand County* makes plain, the overarching thematic principle of the book is the inculcation of the idea—through narrative description, discursive exposition, abstractive generalization, and occasional preachment—"that land is a community" (viii). The community concept is "the basic concept of ecology" (viii). Once land is popularly perceived as a biotic community—as it is professionally perceived in ecology—a correlative land ethic will emerge in the collective cultural consciousness.

Although anticipated as far back as the mid-eighteenth century—in the notion of an "economy of nature"—the concept of the biotic community was more fully and deliberately developed as a working model or paradigm for ecology by Charles Elton in the 1920s.[17] The natural world is organized as an intricate corporate society in which plants and animals occupy "niches," or as Elton alternatively called them, "roles" or "professions," in the economy of nature.[18] As in a feudal community, little or no socioeconomic mobility (upward or otherwise) exists in the biotic community. One is born to one's trade.

Human society, Leopold argues, is founded, in large part, upon mutual security and economic interdependency and preserved only by limitations on freedom of action in the struggle for existence—that is, by ethical constraints. Since the biotic community exhibits, as modern ecology reveals, an analogous structure, it too can be preserved, given the newly amplified impact of "mechanized man," only by analogous limitations on freedom of action—that is, by a land ethic (viii). A land ethic, furthermore, is not only "an ecological necessity," but an "evolutionary possibility" because a moral response to the natural environment—Darwin's social sympathies, sentiments, and instincts translated and codified into a body of principles and precepts—would be automatically triggered in human beings by ecology's social representation of nature (203).

Therefore, the key to the emergence of a land ethic is, simply, universal ecological literacy.

The land ethic rests upon three scientific cornerstones: (1) evolutionary and (2) ecological biology set in a background of (3) Copernican astronomy. Evolutionary theory provides the conceptual link between ethics and social organization and development. It provides a sense of "kinship with fellow-creatures" as well, "fellow-voyagers" with us in the "odyssey of evolution" (109). It establishes a diachronic link between people and non-human nature.

Ecological theory provides a synchronic link—the community concept—a sense of social integration of human and nonhuman nature. Human beings, plants, animals, soils, and waters are "all interlocked in one humming community of cooperations and competitions, one biota."[19] The simplest reason, to paraphrase Darwin, should, therefore, tell each individual that he or she ought to extend his or her social instincts and sympathies to all the members of the biotic community though different from him or her in appearance or habits.

And although Leopold never directly mentions it in *A Sand County Almanac,* the Copernican perspective, the perception of the Earth as "a small planet" in an immense and utterly hostile universe beyond, contributes, perhaps subconsciously, but nevertheless very powerfully, to our sense of kinship, community, and interdependence with fellow denizens of the Earth household. It scales the Earth down to something like a cozy island paradise in a desert ocean.

Here in outline, then, are the conceptual and logical foundations of the land ethic: Its conceptual elements are a Copernican cosmology, a Darwinian protosociobiological natural history of ethics, Darwinian ties of kinship among all forms of life on Earth, and an Eltonian model of the structure of biocenoses all overlaid on a Humean-Smithian moral psychology. Its logic is that natural selection has endowed human beings with an affective moral response to perceived bonds of kinship and community membership and identity; that today the natural environment, the land, is represented as a community, the biotic community; and that, therefore, an environmental or land ethic is both possible—the biopsychological and cognitive conditions are in place—and necessary, since human beings collectively have acquired the power to destroy the integrity, diversity, and stability of the environing and supporting economy of nature. In the remainder of this essay I discuss special features and problems of the land ethic germane to moral philosophy.

J. BAIRD CALLICOTT

The most salient feature of Leopold's land ethic is its provision of what Kenneth Goodpaster has carefully called "moral considerability" for the biotic community per se, not just for fellow members of the biotic community:[20]

> In short, a land ethic changes the role of *Homo sapiens* from conqueror of the land-community to plain member and citizen of it. It implies respect for his fellow-members, *and also respect for the community as such.* (204, emphasis added)

The land ethic, thus, has a holistic as well as an individualistic cast.

Indeed, as "The Land Ethic" develops, the focus of moral concern shifts gradually away from plants, animals, soils, and waters severally to the biotic community collectively. Toward the middle, in the subsection called Substitutes for a Land Ethic, Leopold invokes the "biotic rights" of *species*—as the context indicates—of wildflowers, songbirds, and predators. In The Outlook, the climactic section of "The Land Ethic," nonhuman natural entities, first appearing as fellow members, then considered in profile as species, are not so much as mentioned in what might be called the "summary moral maxim" of the land ethic: "A thing is right when it tends to preserve the integrity, stability, and beauty of the biotic community. It is wrong when it tends otherwise" (224–25).

By this measure of right and wrong, not only would it be wrong for a farmer, in the interest of higher profits, to clear the woods off a 75 percent slope, turn his cows into the clearing, and dump its rainfall, rocks, and soil into the community creek, it would also be wrong for the federal fish and wildlife agency, in the interest of individual animal welfare, to permit populations of deer, rabbits, feral burros, or whatever to increase unchecked and thus to threaten the integrity, stability, and beauty of the biotic communities of which they are members. The land ethic not only provides moral considerability for the biotic community per se, but ethical consideration of its individual members is preempted by concern for the preservation of the integrity, stability, and beauty of the biotic community. The land ethic, thus, not only has a holistic aspect; it is holistic with a vengeance.

The Conceptual Foundations of the Land Ethic

The holism of the land ethic, more than any other feature, sets it apart from the predominant paradigm of modern moral philosophy. It is, therefore, the feature of the land ethic which requires the most patient theoretical analysis and the most sensitive practical interpretation.

As Kenneth Goodpaster pointed out, mainstream modern ethical philosophy has taken egoism as its point of departure and reached a wider circle of moral entitlement by a process of generalization: [21] I am sure that *I*, the enveloped ego, am intrinsically or inherently valuable and thus that *my* interests ought to be considered, taken into account, by "others" when their actions may substantively affect *me*. My own claim to moral consideration, according to the conventional wisdom, ultimately rests upon a psychological capacity—rationality or sentiency were the classical candidates of Kant and Bentham, respectively—which is arguably valuable in itself and which thus qualifies *me* for moral standing.[22] However, then I am forced grudgingly to grant the same moral consideration I demand from others, on this basis, to those others who can also claim to possess the same general psychological characteristic.

A *criterion* of moral value and consideration is thus identified. Goodpaster convincingly argues that mainstream modern moral theory is based, when all the learned dust has settled, on this simple paradigm of ethical justification and logic exemplified by the Benthamic and Kantian prototypes.[23] If the criterion of moral values and consideration is pitched low enough—as it is in Bentham's criterion of sentiency—a wide variety of animals are admitted to moral entitlement.[24] If the criterion of moral value and consideration is pushed lower still—as it is in Albert Schweitzer's reverence-for-life ethic—all minimally conative things (plants as well as animals) would be extended moral considerability.[25] The contemporary animal liberation/rights, and reverence-for-life/life-principle ethics are, at bottom, simply direct applications of the modern classical paradigm of moral argument. But this standard modern model of ethical theory provides no possibility whatever for the moral consideration of wholes—of threatened *populations* of animals and plants, or of endemic, rare, or endangered *species*, or of biotic *communities*, or most expansively, of the *biosphere* in its totality—since wholes per se have no psychological experience of any kind.[26] Because mainstream modern moral theory has been "psychocentric," it has been radically and intractably individualistic or "atomistic" in its fundamental theoretical orientation.

Hume, Smith, and Darwin diverged from the prevailing theoretical model by recognizing that altruism is as fundamental and autochthonous in human nature as is egoism. According to their analysis, moral value is not identified with a natural quality objectively present in morally considerable beings—as reason and/ or sentiency is objectively present in people and/or animals—it is, as it were, projected by valuing subjects.[27]

Hume and Darwin, furthermore, recognize inborn moral sentiments which have society as such as their natural object. Hume insists that "we must renounce the theory which accounts for every moral sentiment by the principle of self-love. We must adopt a more *public affection* and allow that the *interests of society* are not, *even on their own account,* entirely indifferent to us."[28] And Darwin, somewhat ironically (since "Darwinian evolution" very often means natural selection operating exclusively with respect to individuals), sometimes writes as if morality had no other object than the commonweal, the welfare of the community as a corporate entity:

> We have now seen that actions are regarded by savages, and were probably so regarded by primeval man, as good or bad, solely as they obviously affect the welfare of the tribe,—not that of the species, nor that of the individual member of the tribe. This conclusion agrees well with the belief that the so-called moral sense is aboriginally derived from social instincts, for both relate at first exclusively to the community.[29]

Theoretically then, the biotic community owns what Leopold, in the lead paragraph of The Outlook, calls "value in the philosophical sense"—i.e., direct moral considerability—because it is a newly discovered proper object of a specially evolved "public affection" or "moral sense" which all psychologically normal human beings have inherited from a long line of ancestral social primates (223).[30]

In the land ethic, as in all earlier stages of social-ethical evolution, there exists a tension between the good of the community as a whole and the "rights" of its individual members considered severally. While The Ethical Sequence section of "The Land Ethic" clearly evokes Darwin's classical biosocial account of the origin and extension of morals, Leopold is actually more explicitly concerned, in that section, with the interplay between the holistic and individualistic moral sentiments—between sympathy and fellow-feeling on the one hand, and public affection for the commonweal on the other:

> The first ethics dealt with the relation between individuals; the Mosaic Decalogue is an example. Later accretions dealt with the relation between the individual and society. The Golden Rule tries to integrate the individual to society; democracy to integrate social organization to the individual. (202–3)

Actually, it is doubtful that the first ethics dealt with the relation between individuals and not at all with the relation between the individual and society. (This, along with the remark that ethics replaced an "original free-for-all competition," suggests that Leopold's Darwinian line of thought has been uncritically tainted with Hobbesean elements. [202]. Of course, Hobbes's "state of nature," in which there prevailed a war of each against all, is absurd from an evolutionary point of view.) A century of ethnographic studies seems to confirm, rather, Darwin's conjecture that the relative weight of the holistic component is greater in tribal ethics—the tribal ethic of the Hebrews recorded in the Old Testament constitutes a vivid case in point—than in more recent accretions. The Golden Rule, on the other hand, does not mention, in any of its formulations, society per se. Rather, its primary concern seems to be "others," i.e., other human individuals. Democracy, with its stress on individual liberties and rights, seems to further rather than countervail the individualistic thrust of the Golden Rule.

In any case, the conceptual foundations of the land ethic provide a well-formed, self-consistent theoretical basis for including both fellow members of the biotic community and the biotic community itself (considered as a corporate entity) within the purview of morals. The preemptive emphasis, however, on the welfare of the community as a whole, in Leopold's articulation of the land ethic, while certainly *consistent* with its Humean-Darwinian theoretical foundations, is not *determined* by them alone. The overriding holism of the land ethic results, rather, more from the way our moral sensibilities are informed by ecology.

Ecological thought, historically, has tended to be holistic in outlook.[31] Ecology is the study of the *relationships* of organisms to one another and to the elemental environment. These relationships bind the *relata*—plants, animals, soils, and waters—into a seamless fabric. The ontological primacy of objects and the ontological subordination of relationships, characteristic of classical Western science, is, in fact, reversed in ecology.[32] Ecological relationships determine the nature of organisms rather than the other way around. A species is what it is because it has adapted to a niche in the ecosystem. The whole, the system itself, thus, literally and quite straightforwardly shapes and forms its component parts.

Antedating Charles Elton's community model of ecology was F. E. Clements' and S. A. Forbes' organism model.[33] Plants and animals, soils and waters, according to this paradigm, are integrated into one superorganism. Species are, as it were, its organs; specimens its cells. Although Elton's community paradigm (later modified, as we shall see, by Arthur Tansley's ecosystem idea) is the principal and morally fertile ecological concept of "The Land Ethic," the more radically holistic superorganism paradigm of Clements and Forbes resonates in "The Land Ethic" as an audible overtone. In the peroration of Land Health and the A-B Cleavage, for example, which immediately precedes The Outlook, Leopold insists that

> in all of these cleavages, we see repeated the same basic paradoxes: man the conqueror *versus* man the biotic citizen; science the sharp-

ener of his sword *versus* science the searchlight on his universe; land the slave and servant *versus* land the collective organism. (223)

And on more than one occasion Leopold, in the latter quarter of "The Land Ethic," talks about the "health" and "disease" of the land—terms which are at once descriptive and normative and which, taken literally, characterize only organisms proper.

In an early essay, "Some Fundamentals of Conservation in the Southwest," Leopold speculatively flirted with the intensely holistic superorganism model of the environment as a paradigm pregnant with moral implications:

> It is at least not impossible to regard the earth's parts—soil, mountains, rivers, atmosphere, etc.—as organs or parts of organs, of *a coordinated whole,* each part with a definite function. And if we could see *this whole, as a whole,* through a great period of time, we might perceive not only organs with coordinated functions, but possibly also that process of consumption and replacement which in biology we call metabolism, or growth. In such a case we would have all the visible attributes of a living thing, which we do not realize to be such because it is too big, and its life processes too slow. And there would also follow that invisible attribute—a soul or consciousness—which . . . many philosophers of all ages ascribe to all living things and aggregates thereof, including the "dead" earth.
>
> Possibly in our intuitive perceptions, which may be truer than our science and less impeded by words than our philosophies, we realize the indivisibility of the earth—its soil, mountains, rivers, forests, climate, plants, and animals—and *respect it collectively* not only as a useful servant but as a living being, vastly less alive than ourselves, but vastly greater than ourselves in time and space. . . . Philosophy, then, suggests one reason why we cannot destroy the earth with moral impunity; namely, that the "dead" earth is an organism possessing a certain kind and degree of life, which we intuitively respect as such.[34]

Had Leopold retained this overall theoretical approach in "The Land Ethic," the land ethic would doubtless have enjoyed more critical attention from philosophers. The moral foundations of a land or, as he might then have called it, "earth" ethic, would rest upon the hypothesis that the Earth is alive and ensouled—possessing inherent psychological characteristics, logically parallel to reason and sentiency. This notion of a conative whole Earth could plausibly have served as a general criterion of intrinsic worth and moral considerability, in the familiar format of mainstream moral thought.

Part of the reason, therefore, that "The Land Ethic" emphasizes more and more the integrity, stability, and beauty of the environment as a whole, and less and less the "biotic right" of individual plants and animals to life, liberty, and the pursuit of happiness, is that the superorganism ecological paradigm invites one, much more than does the community paradigm, to hypostatize, to reify the whole, and to subordinate its individual members.

In any case, as we see, rereading "The Land Ethic" in light of "Some Fundamentals," the whole Earth organism image of nature is vestigially present in Leopold's later thinking. Leopold may have abandoned the "earth ethic" because ecology had abandoned the organism analogy, in favor of the community analogy, as a working theoretical paradigm. And the community model was more suitably given moral implications by the social/sentimental ethical natural history of Hume and Darwin.

Meanwhile, the biotic community ecological paradigm itself had acquired, by the late thirties and forties, a more holistic cast of its own. In 1935 British ecologist Arthur Tansley pointed out that from the perspective of physics the "currency" of the "economy of nature" is energy.[35] Tansley suggested that Elton's qualitative and descriptive food chains, food webs, trophic niches, and biosocial professions could be quantitatively expressed by means of a thermodynamic flow model. It is Tansley's state-of-the-art thermodynamic paradigm of the environment that Leopold explicitly sets out as a "mental image of land" in relation to which "we can be ethical" (214). And it is the ecosystemic model of land which informs the cardinal practical precepts of the land ethic.

The Land Pyramid is the pivotal section of "The Land Ethic"—the section which effects a complete transition from concern for "fellow-members" to the "community as such." It is also its longest and most technical section. A description of the "ecosystem" (Tansley's deliberately nonmetaphorical term) begins with the

The Conceptual Foundations of the Land Ethic

sun. Solar energy "flows through a circuit called the biota" (215). It enters the biota through the leaves of green plants and courses through plant-eating animals, and then on to omnivores and carnivores. At last the tiny fraction of solar energy converted to biomass by green plants remaining in the corpse of a predator, animal feces, plant detritus, or other dead organic material is garnered by decomposers—worms, fungi, and bacteria. They recycle the participating elements and degrade into entropic equilibrium any remaining energy. According to this paradigm

> land, then, is not merely soil; it is a fountain of energy flowing through a circuit of soils, plants, and animals. Food chains are the living channels which conduct energy upward; death and decay return it to the soil. The circuit is not closed; . . . but it is a sustained circuit, like a slowly augmented revolving fund of life. (216)

In this exceedingly abstract (albeit poetically expressed) model of nature, process precedes substance and energy is more fundamental than matter. Individual plants and animals become less autonomous beings than ephemeral structures in a patterned flux of energy. According to Yale biophysicist Harold Morowitz,

> viewed from the point of view of modern [ecology], each living thing . . . is a dissipative structure, that is it does not endure in and of itself but only as a result of the continual flow of energy in the system. An example might be instructive. Consider a vortex in a stream of flowing water. The vortex is a structure made of an ever-changing group of water molecules. It does not exist as an entity in the classical Western sense; it exists only because of the flow of water through the stream. In the same sense, the structures out of which biological entities are made are transient, unstable entities with constantly changing molecules, dependent on a constant flow of energy from food in order to maintain form and structure. . . . From this point of view the reality of individuals is problematic because they do not exist per se but only as local perturbations in this universal energy flow.[36]

Though less bluntly stated and made more palatable by the unfailing charm of his prose, Leopold's proffered mental image of land is just as expansive, systemic, and distanced as Morowitz'. The maintenance of "the complex structure of the land and its smooth functioning as an energy unit" emerges in The Land Pyramid as the *summum bonum* of the land ethic (216).

From this good Leopold derives several practical principles slightly less general, and therefore more substantive, than the summary moral maxim of the land ethic distilled in The Outlook. "The trend of evolution [not its "goal," since evolution is ateleological] is to elaborate and diversify the biota" (216). Hence, among our cardinal duties is the duty to preserve what species we can, especially those at the apex of the pyramid—the top carnivores. "In the beginning, the pyramid of life was low and squat; the food chains short and simple. Evolution has added layer after layer, link after link" (215–16). Human activities today, especially those, like systematic deforestation in the tropics, resulting in abrupt massive extinctions of species, are in effect "devolutionary"; they flatten the biotic pyramid; they choke off some of the channels and gorge others (those which terminate in our own species).[37]

The land ethic does not enshrine the ecological status quo and devalue the dynamic dimension of nature. Leopold explains that "evolution is a long series of self-induced changes, the net result of which has been to elaborate the flow mechanism and to lengthen the circuit. Evolutionary changes, however, are usually slow and local. Man's invention of tools has enabled him to make changes of unprecedented violence, rapidity, and scope" (216–17). "Natural" species extinction, i.e., species extinction in the normal course of evolution, occurs when a species is replaced by competitive exclusion or evolves into another form.[38] Normally speciation outpaces extinction. Mankind inherited a richer, more diverse world than had ever existed before in the 3.5 billion-year odyssey of life on Earth.[39] What is wrong with anthropogenic species extirpation and extinction is the *rate* at which it is occurring and the *result:* biological impoverishment instead of enrichment.

Leopold goes on here to condemn, in terms of its impact on the ecosystem, "the world-wide pooling of faunas and floras," i.e., the indiscriminate introduction of exotic and domestic species and the dislocation of native and endemic species; mining the soil for its stored biotic energy, leading ultimately to diminished fertility and to erosion; and polluting and damming water courses (217).

According to the land ethic, therefore: Thou shalt not extirpate or render species extinct; thou shalt exercise great caution in introducing exotic and domestic species into local ecosystems, in extracting energy from the soil and releasing it into the biota, and in damming or polluting water courses; and thou shalt be especially solicitous of predatory birds and mammals. Here in brief are the express moral precepts of the land ethic. They are all explicitly informed—not to say derived—from the energy circuit model of the environment.

The living channels—"food chains"—through which energy courses are composed of individual plants and animals. A central, stark fact lies at the heart of ecological processes: Energy, the currency of the economy nature, passes from one organism to another, not from hand to hand, like coined money, but, so to speak, from stomach to stomach. Eating *and being eaten*, living *and dying* are what make the biotic community hum.

The precepts of the land ethic, like those of all previous accretions, reflect and reinforce the structure of the community to which it is correlative. Trophic asymmetries constitute the kernel of the biotic community. It seems unjust, unfair. But that is how the economy of nature is organized (and has been for thousands of millions of years). The land ethic, thus, affirms as good, and strives to preserve, the very inequities in nature whose social counterparts in human communities are condemned as bad and would be eradicated by familiar social ethics, especially by the more recent Christian and secular egalitarian exemplars. A "right to life" for individual members is not consistent with the structure of the biotic community and hence is not mandated by the land ethic. This disparity between the land ethic and its more familiar social precedents contributes to the apparent devaluation of individual *members* of the biotic community and augments and reinforces the tendency of the land ethic, driven by the systemic vision of ecology, toward a more holistic or community-per-se orientation.

Of the few moral philosophers who have given the land ethic a moment's serious thought, most have regarded it with horror because of its emphasis on the good of the community and its deemphasis on the welfare of individual members of the community. Not only are other sentient creatures members of the biotic community and subordinate to its integrity, beauty, and stability; so are *we*. Thus, if it is not only morally permissible, from the point of view of the land ethic, but morally required, that members of certain species be abandoned to predation and other vicissitudes of wild life or even deliberately culled (as in the case of alert and sentient whitetail deer) for the sake of the integrity, stability, and beauty of the biotic community, how can we consistently exempt ourselves from a similar draconian regime? We too are only "plain members and citizens" of the biotic community. And our global population is growing unchecked. According to William Aiken, from the point of view of the land ethic, therefore, "massive human diebacks would be good. It is our duty to cause them. It is our species' duty, relative to the whole, to eliminate 90 percent of our numbers." Thus, according to Tom Regan, the land ethic is a clear case of "environmental fascism."[40]

Of course Leopold never intended the land ethic to have either inhumane or antihumanitarian implications or consequences. But whether he intended them or not, a logically consistent deduction from the theoretical premises of the land ethic might force such untoward conclusions. And given their magnitude and monstrosity, these derivations would constitute a *reductio ad absurdum* of the whole land ethic enterprise and entrench and reinforce our current human chauvinism and moral alienation from nature. If this is what membership in the biotic community entails, then all but the most radical misanthropes would surely want to opt out.

The land ethic, happily, implies neither inhumane nor inhuman consequences. That some philosophers think it must follows more from their own theoretical presuppositions than from the theoretical elements of the land ethic itself. Conventional modern ethical theory rests moral entitlement, as I earlier pointed out, on a criterion or qualification. If a candidate meets the criterion—rationality or sentiency are the most commonly posited—he, she, or it is entitled to equal moral standing with others who possess the same qualification in equal degree. Hence, reasoning in this philosophically orthodox way, and forcing Leopold's theory to conform: if human beings are, with other animals, plants, soils, and waters, equally members of the biotic community, and if community membership is the criterion of equal moral consideration, then not only do animals, plants, soils, and waters have equal (highly attenuated) "rights," but human beings are equally subject to the same subordination of individual welfare and rights in respect to the good of the community as a whole.

But the land ethic, as I have been at pains to point out, is heir to a line of moral analysis different from that institutionalized in contemporary moral philosophy. From the biosocial evolutionary analysis of ethics upon which Leopold builds the land ethic, it (the land ethic) neither replaces nor overrides previous accretions. Prior moral sensibilities and obligations attendant upon and correlative to prior strata of social involvement remain operative and preemptive.

Being citizens of the United States, or the United Kingdom, or the Soviet Union, or Venezuela, or some other nation-state, and therefore having national obligations and patriotic duties, does not mean that we are not also members of smaller communities or social groups—cities or townships, neighborhoods, and families—or that we are relieved of the peculiar moral responsibilities attendant upon and correlative to these memberships as well. Similarly, our recognition of the biotic community and our immersion in it does not imply that we do not also remain members of the human community—the "family of man" or "global village"—or that we are relieved of the attendant and correlative moral responsibilities of that membership, among them to respect universal human rights and uphold the principles of individual human worth and dignity. The biosocial development of morality does not grow in extent like an expanding balloon, leaving no trace of its previous boundaries, so much as like the circumference of a tree.[41] Each emergent, and larger, social unit is layered over the more primitive, and intimate, ones.

Moreover, as a general rule, the duties correlative to the inner social circles to which we belong eclipse those correlative to the rings farther from the heartwood when conflicts arise. Consider our moral revulsion when zealous ideological nationalists encourage children to turn their parents in to the authorities if their parents should dissent from the political or economic doctrines of the ruling party. A zealous environmentalist who advocated visiting war, famine, or pestilence on human populations (those existing somewhere else, of course) in the name of the integrity, beauty, and stability of the biotic community would be similarly perverse. Family obligations in general come before nationalistic duties and humanitarian obligations in general come before environmental duties. The land ethic, therefore, is not draconian or fascist. It does not cancel human morality. The land ethic may, however, as with any new accretion, demand choices which affect, in turn, the demands of the more interior social-ethical circles. Taxes and the military draft may conflict with family-level obligations. While the land ethic, certainly, does not cancel human morality, neither does it leave it unaffected.

Nor is the land ethic inhumane. Nonhuman fellow members of the biotic community have no "human rights," because they are not, by definition, members of the human community. As fellow members of the biotic community, however, they deserve respect.

How exactly to express or manifest respect, while at the same time abandoning our fellow members of the biotic community to their several fates or even actively consuming them for our own needs (and wants), or deliberately making them casualties of wildlife management for ecological integrity, is a difficult and delicate question.

Fortunately, American Indian and other traditional patterns of human-nature interaction provide rich and detailed models. Algonkian woodland peoples, for instance, represented animals, plants, birds, waters, and minerals as other-than-human persons engaged in reciprocal, mutually beneficial socioeconomic intercourse with human beings.[42] Tokens of payment, together with expressions of apology, were routinely offered to the beings whom it was necessary for these Indians to exploit. Care not to

The Conceptual Foundations of the Land Ethic

waste the usable parts, and care in the disposal of unusable animal and plant remains, were also an aspect of the respectful, albeit necessarily consumptive, Algonkian relationship with fellow members of the land community. As I have more fully argued elsewhere, the Algonkian portrayal of human-nature relationships is, indeed, although certainly different in specifics, identical in abstract form to that recommended by Leopold in the land ethic.[43]

Ernest Partridge has turned the existence of an American Indian land ethic, however, against the historicity of the biosocial theoretical foundations of the land ethic:

> Anthropologists will find much to criticize in [Leopold's] account.
> . . . The anthropologist will point out that in many primitive cultures, far greater moral concern may be given to animals or even to trees, rocks, and mountains, than are given to persons in other tribes. . . . Thus we find not an "extension of ethics," but a "leapfrogging" of ethics, over and beyond persons to natural beings and objects. Worse still for Leopold's view, a primitive culture's moral concern for nature often appears to "draw back" to a human centered perspective as that culture evolves toward a civilized condition.[44]

Actually, the apparent historical anomalies, which Partridge points out, confirm, rather than confute, Leopold's ethical sequence. At the tribal stage of human social evolution, a member of another tribe was a member of a separate and independent social organization, and hence of a separate and alien moral community; thus, "[human] persons in other tribes" were not extended moral consideration, just as the biosocial model predicts. However, at least among those tribal people whose world view I have studied in detail, the animals, trees, rocks, and mountains of a tribe's territory were portrayed as working members and trading partners of the local community. Totem representation of clan units within tribal communities facilitated this view. Groups of people were identified as cranes, bears, turtles, and so on; similarly, populations of deer, beaver, fox, etc., were clans of "people"—people who liked going about in outlandish get-ups. Frequent episodes in tribal mythologies of "metamorphosis"—

the change from animal to human form and vice versa—further cemented the tribal integration of local nonhuman natural entities. It would be very interesting to know if the flora and fauna living in another tribe's territory would be regarded, like its human members, as beyond the moral pale.

Neither does the "'draw-back' to a human centered [ethical] perspective as [a] culture evolves toward a civilized condition," noticed by Partridge, undermine the biosocial theoretical foundations of the land ethic. Rather, the biosocial theoretical foundations of the land ethic elucidate this historical phenomenon as well. As a culture evolves toward civilization, it increasingly distances itself from the biotic community. "*Civil*ization" means "cityfication"—inhabitation of and participation in an artificial, humanized environment and a corresponding perception of isolation and alienation from nature. Nonhuman natural entities, thus, are divested of their status as members in good standing of the moral community as civilization develops. Today, two processes internal to civilization are bringing us to a recognition that our renunciation of our biotic citizenship was a mistaken self-deception. Evolutionary science and ecological science, which certainly are products of modern civilization now supplanting the anthropomorphic and anthropocentric myths of earlier civilized generations, have rediscovered our integration with the biotic community. And the negative feedback received from modern civilization's technological impact upon nature— pollution, biological impoverishment, etc.—forcefully reminds us that mankind never really has, despite past assumptions to the contrary, existed apart from the environing biotic community.

This reminder of our recent rediscovery of our biotic citizenship brings us face to face with the paradox posed by Peter Fritzell:[45] Either we are plain members and citizens of the biotic community, on a par with other creatures, or we are not. If we are, then we have no moral obligations to our fellow members or to the community per se because, as understood from a modern scientific perspective, nature and natural phenomena are amoral. Wolves and alligators do no wrong in killing and eating deer and dogs (respectively). Elephants cannot be blamed for bulldozing

acacia trees and generally wreaking havoc in their natural habitats. If human beings are natural beings, then human behavior,
however destructive, is natural behavior and is as blameless, from
a natural point of view, as any other behavioral phenomenon exhibited by other natural beings. On the other hand, we are moral
beings, the implication seems clear, precisely to the extent that
we are civilized, that we have removed ourselves from nature. We
are more than natural beings; we are metanatural—not to say,
"supernatural"—beings. But then our moral community is limited to only those beings who share our transcendence of nature,
i.e., to human beings (and perhaps to pets who have joined our
civilized community as surrogate persons) and to the human
community. Hence, have it either way—we are members of the
biotic community or we are not—a land or environmental ethic
is aborted by either choice.

But nature is *not* amoral. The tacit assumption that we are de-
liberating, choice-making ethical beings only to the extent that
we are metanatural, civilized beings, generates this dilemma. The
biosocial analysis of human moral behavior, in which the land
ethic is grounded, is designed precisely to show that in fact intelligent moral behavior *is* natural behavior. Hence, we are moral
beings not in spite of, but in accordance with, nature. To the extent that nature has produced at least one ethical species, *Homo
sapiens,* nature is not amoral.

Alligators, wolves, and elephants are not subject to reciprocal
interspecies duties or land ethical obligations themselves because
they are incapable of conceiving and/or assuming them. Alligators, as mostly solitary, entrepreneurial reptiles, have no apparent moral sentiments or social instincts whatever. And while
wolves and elephants certainly do have social instincts and at
least protomoral sentiments, as their social behavior amply indicates, their conception or imagination of community appears to
be less culturally plastic than ours and less amenable to cognitive
information. Thus, while we might regard them as ethical beings, they are not able, as we are, to form the concept of a universal biotic community, and hence conceive an all-inclusive, holistic land ethic.

The paradox of the land ethic, elaborately noticed by Fritzell, may be cast more generally still in more conventional philosophical terms: Is the land ethic prudential or deontological? Is the land ethic, in other words, a matter of enlightened (collective, human) self-interest, or does it genuinely admit nonhuman natural entities and nature as a whole to true moral standing?

The conceptual foundations of the land ethic, as I have here set them out, and much of Leopold's hortatory rhetoric, would certainly indicate that the land ethic is deontological (or duty oriented) rather than prudential. In the section significantly titled The Ecological Conscience, Leopold complains that the then-current conservation philosophy is inadequate because "it defines no right or wrong, assigns no obligation, calls for no sacrifice, implies no change in the current philosophy of values. In respect of land-use, it urges *only* enlightened self-interest" (207–8, emphasis added). Clearly, Leopold himself thinks that the land ethic goes beyond prudence. In this section he disparages mere "self-interest" two more times, and concludes that "obligations have no meaning without conscience, and the problem we face is the extension of the social conscience from people to land" (209).

In the next section, Substitutes for a Land Ethic, he mentions rights twice—the "biotic right" of birds to continuance and the absence of a right on the part of human special interest to exterminate predators.

Finally, the first sentences of The Outlook read: "It is inconceivable to me that an ethical relation to land can exist without love, respect, and admiration for land, and a high regard for its value. By value, I of course mean something far broader than mere economic value; I mean value in the philosophical sense" (223). By "value in the philosophical sense," Leopold can only mean what philosophers more technically call "intrinsic value" or "inherent worth." [46] Something that has intrinsic value or inherent worth is valuable in and of itself, not because of what it can do for us. "Obligation," "sacrifice," "conscience," "respect," the ascription of rights, and intrinsic value—all of these are consistently opposed to self-interest and seem to indicate decisively that the land ethic is of the deontological type.

Some philosophers, however, have seen it differently. Scott Lehmann, for example, writes,

The Conceptual Foundations of the Land Ethic

Although Leopold claims for communities of plants and animals a "right to continued existence," his argument is homocentric, appealing to the human stake in preservation. Basically it is an argument from enlightened self-interest, where the self in question is not an individual human being but humanity—present and future—as a whole. . . .[47]

Lehmann's claim has some merits, even though it flies in the face of Leopold's express commitments. Leopold does frequently lapse into the language of (collective, long-range, human) self-interest. Early on, for example, he remarks, "in human history, we have learned (I hope) that the conqueror role is eventually *self*-defeating" (204, emphasis added). And later, of the 95 percent of Wisconsin's species which cannot be "sold, fed, eaten, or otherwise put to economic use," Leopold reminds us that "these creatures are members of the biotic community, and if (as I believe) its stability depends on its integrity, they are entitled to continuance" (210). The implication is clear: the economic 5 percent cannot survive if a significant portion of the uneconomic 95 percent are extirpated; nor may *we*, it goes without saying, survive without these "resources."

Leopold, in fact, seems to be consciously aware of this moral paradox. Consistent with the biosocial foundations of his theory, he expresses it in sociobiological terms:

An ethic may be regarded as a mode of guidance for meeting ecological situations so new or intricate, or involving such deferred reactions, that the path of social expediency is not discernible to the average individual. Animal instincts are modes of guidance for the individual in meeting such situations. Ethics are possibly a kind of community instinct in-the-making. (203)

From an objective, descriptive sociobiological point of view, ethics evolve because they contribute to the inclusive fitness of their carriers (or, more reductively still, to the multiplication of their carriers' genes); they are expedient. However, the path to self-interest (or to the self-interest of the selfish gene) is not discernible to the participating individuals (nor, certainly, to their genes). Hence, ethics are grounded in instinctive feeling—love, sympathy, respect—not in self-conscious calculating intelligence. Somewhat like the paradox of hedonism—the notion that one cannot achieve happiness if one directly pursues happiness per se and not other things—one can only secure self-interest by putting the interests of others on a par with one's own (in this case long-range collective human self-interest and the interest of other forms of life and of the biotic community per se).

So, is the land ethic deontological or prudential, after all? It is both—self-consistently both—depending upon point of view. From the inside, from the lived, felt point of view of the community member with evolved moral sensibilities, it is deontological. It involves an affective-cognitive posture of genuine love, respect, admiration, obligation, self-sacrifice, conscience, duty, and the ascription of intrinsic value and biotic rights. From the outside, from the objective and analytic scientific point of view, it is prudential. "There is no other way for land to survive the impact of mechanized man," nor, therefore, for mechanized man to survive his own impact upon the land (viii).

Notes

1. Wallace Stegner, "The Legacy of Aldo Leopold"; Curt Meine, "Building 'The Land Ethic'"; both in this volume. The oft-repeated characterization of Leopold as a prophet appears traceable to Roberts Mann, "Aldo Leopold: Priest and Prophet," *American Forests* 60, no. 8 (August 1954): 23, 42–43; it was picked up, apparently, by Ernest Swift, "Aldo Leopold: Wisconsin's Conservationist Prophet," *Wisconsin Tales and Trails* 2, no. 2 (September 1961): 2–5; Roderick Nash institutionalized it in his chapter, "Aldo Leopold: Prophet," in *Wilderness and the American Mind* (New Haven: Yale University Press, 1967; revised edition, 1982).

2. John Passmore, *Man's Responsibility for* [significantly not *"to"*] *Nature: Ecological Problems and Western Traditions* (New York: Charles Scribner's Sons, 1974).

3. H. J. McCloskey, *Ecological Ethics and Politics* (Totowa, N.J.: Rowman and Littlefield, 1983), 56.

4. Robin Attfield, in "Value in the Wilderness," *Metaphilosophy* 15 (1984), writes, "Leopold the philosopher is something of a disaster, and I dread the thought of the student whose concept of philosophy is modeled principally on these extracts. (Can value 'in the philosophical sense' be contrasted with instrumental value? If concepts of right and wrong did not apply to slaves in Homeric Greece, how could Odysseus

The Conceptual Foundations of the Land Ethic

suspect the slavegirls of 'misbehavior'? If all ethics rest on interdependence how are obligations to infants and small children possible? And how can 'obligations have no meaning without conscience,' granted that the notion of conscience is conceptually dependent on that of obligation?)" (294). L. W. Sumner, "Review of Robin Attfield, *The Ethics of Environmental Concern," Environmental Ethics* 8 (1986): 77.

5. Edward O. Wilson, *Sociobiology: The New Synthesis* (Cambridge: Harvard University Press, 1975), 3. See also W. D. Hamilton, "The Genetical Theory of Social Behavior," *Journal of Theoretical Biology* 7 (1964): 1–52.

6. Charles R. Darwin, *The Descent of Man and Selection in Relation to Sex* (New York: J. A. Hill and Company, 1904). The quoted phrase occurs on p. 97.

7. See Adam Smith, *Theory of the Moral Sentiments* (London and Edinburgh: A. Millar, A. Kinkaid, and J. Bell, 1759) and David Hume, *An Enquiry Concerning the Principles of Morals* (Oxford: The Clarendon Press, 1777; first published in 1751). Darwin cites both works in the key fourth chapter of *Descent* (pp. 106 and 109, respectively).

8. Darwin, *Descent,* 98ff.

9. Ibid., 105f.

10. Ibid., 113ff.

11. Ibid., 105.

12. See, for example, Elman R. Service, *Primitive Social Organization: An Evolutionary Perspective* (New York: Random House, 1962).

13. See Marshall Sahlins, *Stone Age Economics* (Chicago: Aldine Atherton, 1972).

14. Darwin, *Descent,* 111.

15. Ibid., 117ff. The quoted phrase occurs on p. 118.

16. Ibid., 124.

17. See Donald Worster, *Nature's Economy: The Roots of Ecology* (San Francisco: Sierra Club Books, 1977).

18. Charles Elton, *Animal Ecology* (New York: Macmillan, 1927).

19. Aldo Leopold, *Round River* (New York: Oxford University Press, 1953), 148.

20. Kenneth Goodpaster, "On Being Morally Considerable," *Journal of Philosophy* 22 (1978): 308–25. Goodpaster wisely avoids the term *rights,* defined so strictly albeit so variously by philosophers, and used so loosely by nonphilosophers.

21. Kenneth Goodpaster, "From Egoism to Environmentalism" in *Ethics and Problems of the 21st Century,* ed. K. E. Goodpaster and K. M. Sayre (Notre Dame, Ind.: University of Notre Dame Press, 1979), 21–35.

22. See Immanuel Kant, *Foundations of the Metaphysics of Morals* (New York: Bobbs-Merrill, 1959; first published in 1785); and Jeremy Bentham, *An Introduction to the Principles of Morals and Legislation,* new edition (Oxford: The Clarendon Press, 1823).

23. Goodpaster, "Egoism to Environmentalism." Actually Goodpaster regards *Hume* and Kant as the cofountainheads of this sort of moral philosophy. But Hume does not reason in this way. For Hume, the other-oriented sentiments are as primitive as self-love.

24. See Peter Singer, *Animal Liberation: A New Ethics for Our Treatment of Animals* (New York: Avon Books, 1975) for animal liberation; and see Tom Regan, *All That Dwell Therein: Animal Rights and Environmental Ethics* (Berkeley: University of California Press, 1982) for animal rights.

25. See Albert Schweitzer, *Philosophy of Civilization: Civilization and Ethics,* trans. John Naish (London: A. & C. Black, 1923). For a fuller discussion see J. Baird Callicott, "On the Intrinsic Value of Non-human Species," in *The Preservation of Species,* ed. Bryan Norton (Princeton: Princeton University Press, 1986), 138–72.

26. Peter Singer and Tom Regan are both proud of this circumstance and consider it a virtue. See Peter Singer, "Not for Humans Only: The Place of Nonhumans in Environmental Issues" in *Ethics and Problems of the 21st Century,* 191–206; and Tom Regan, "Ethical Vegetarianism and Commercial Animal Farming" in *Contemporary Moral Problems,* ed. James E. White (St. Paul, Minn.: West Publishing Co., 1985), 279–94.

27. See J. Baird Callicott, "Hume's Is/Ought Dichotomy and the Relation of Ecology to Leopold's Land Ethic," *Environmental Ethics* 4 (1982): 163–74, and "Non-anthropocentric Value Theory and Environmental Ethics," *American Philosophical Quarterly* 21 (1984): 299–309, for an elaboration.

28. Hume, *Enquiry,* 219.

29. Darwin, *Descent,* 120.

30. I have elsewhere argued that "value in the philosophical sense" means "intrinsic" or "inherent" value. See J. Baird Callicott, "The Philosophical Value of Wildlife," in *Valuing Wildlife: Economic and Social Values of Wildlife,* ed. Daniel J. Decker and Gary Goff (Boulder, Col.: Westview Press, 1986), 214–221.

31. See Worster, *Nature's Economy.*

32. See J. Baird Callicott, "The Metaphysical Implications of Ecology," *Environmental Ethics* 8 (1986): 300–315, for an elaboration of this point.

33. Robert P. McIntosh, *The Background of Ecology: Concept and Theory* (Cambridge: Cambridge University Press, 1985).

34. Aldo Leopold, "Some Fundamentals of Conservation in the Southwest," *Environmental Ethics* 1 (1979): 139–40, emphasis added.

35. Arthur Tansley, "The Use and Abuse of Vegetational Concepts and Terms," *Ecology* 16 (1935): 292–303.

36. Harold J. Morowitz, "Biology as a Cosmological Science," *Main Currents in Modern Thought* 28 (1972): 156.

37. I borrow the term "devolution" from Austin Meredith, "Devolution," *Journal of Theoretical Biology* 96 (1982): 49–65.

38. Holmes Rolston, III, "Duties to Endangered Species," *Bioscience* 35 (1985): 718–26. See also Geerat Vermeij, "The Biology of Human-Caused Extinction," in Norton, *Preservation of Species*, 28–49.

39. See D. M. Raup and J. J. Sepkoski, Jr., "Mass Extinctions in the Marine Fossil Record," *Science* 215 (1982): 1501–3.

40. William Aiken, "Ethical Issues in Agriculture," in *Earthbound: New Introductory Essays in Environmental Ethics,* ed. Tom Regan (New York: Random House, 1984), 269. Tom Regan, *The Case for Animal Rights* (Berkeley: University of California Press, 1983) 262, and "Ethical Vegetarianism," 291. See also Eliott Sober, "Philosophical Problems for Environmentalism," in Norton, *Preservation of Species*, 173–94.

41. I owe the tree-ring analogy to Richard and Val Routley (now Sylvan and Plumwood, respectively), "Human Chauvinism and Environmental Ethics," in *Environmental Philosophy,* ed. D. Mannison, M. McRobbie, and R. Routley (Canberra: Department of Philosophy, Research School of the Social Sciences, Australian National University, 1980), 96–189. A good illustration of the balloon analogy may be found in Peter Singer, *The Expanding Circle: Ethics and Sociobiology* (New York: Farrar, Straus and Giroux, 1983).

42. For an elaboration see Thomas W. Overholt and J. Baird Callicott, *Clothed-in-Fur and Other Tales: An Introduction to an Ojibwa World View* (Washington, D.C.: University Press of America, 1982).

43. J. Baird Callicott, "Traditional American Indian and Western European Attitudes Toward Nature: An Overview," *Environmental Ethics* 4 (1982): 163–74.

44. Ernest Partridge, "Are We Ready for an Ecological Morality?" *Environmental Ethics* 4 (1982): 177.

45. Peter Fritzell, "The Conflicts of Ecological Conscience," in this volume.

46. See Worster, *Nature's Economy.*

47. Scott Lehmann, "Do Wildernesses Have Rights?" *Environmental Ethics* 3 (1981): 131.

IV * The Impact

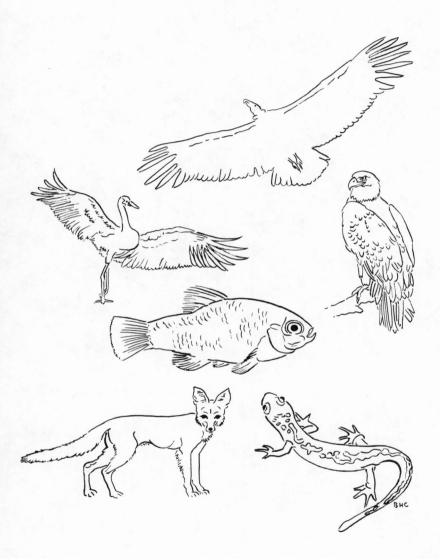

10 ✳ A Pilgrim's Progress from Group A to Group B

EDWIN P. PISTER

"Conservationists are notorious for their dissensions. . . . In each field one group (A) regards the land as soil, and its function as commodity-production; another group (B) regards the land as a biota, and its function as something broader" (221).

"Conservation is a state of harmony between men and land" (207).

"Rest! cries the chief sawyer, and we pause for breath." My eyes fell upon the brass plaque, recently fastened to a large boulder near where the "good oak" had once stood (not far from "the shack"), and waves of nostalgia and emotion washed over me as my mind quickly retraced the events that led me here, and the role played by *A Sand County Almanac* in my evolution as a professional steward of natural resources. I remembered vividly that spring day in Berkeley, thirty-six years earlier, when A. Starker Leopold suggested that his undergraduate students in Wildlife Management read "this group of essays that Dad wrote." In 1949, I was unable to absorb much of Aldo Leopold's philosophy of conservation. It first had to be tempered by more than a decade of experience and exposure in the field.

Following the usual variety of moves, jobs, and agencies which accompany the earlier portion of most careers in fish and wildlife research and management, I settled, in the late 1950s, into a position as a fishery biologist with the California Department of Fish and Game in the eastern Sierra and desert regions of the state. My graduate studies in limnology had already given me considerable knowledge of the geography of the area and an insight into the biological characteristics of many of the waters falling under my jurisdiction. With but one assistant, I was relegated the responsibility for the management of nearly a thousand waters extending from the crest of the Sierra Nevada eastward to the Nevada state line, and ranging from the top of Mt. Whitney at 4,418 meters, to the floor of Death Valley at nearly 100 meters below sea level. "Management," ideally, meant responsibility for the perpetuation of *all* species of aquatic organisms—including fishes, amphibians, invertebrates, and even reptiles—and their habitats.

The management programs which I had inherited from my predecessors reflected the philosophies of the times. They were, technologically, "state of the art," and they were designed to meet the desires and demands of a public hungry for outdoor recreation following World War II. They were, in short, model "utilitarian" management procedures.

Fleets of tanker trucks from huge and highly efficient trout hatcheries did a superb job of meeting angler demand, despite the obvious fact that each planted trout was an "artificialized" trophy. However, the program sold licenses (the department's primary income source), which increased the funding of the program, which grew ever larger. And the larger it grew, the greater became the bureaucratic intransigence. It was popular, it kept us fully employed, it fueled the tourist economy, and it was heresy to think otherwise.

Most California fishery biologists scurried about, robotlike, in a heroic effort to increase catch per angler effort. With one exception, the golden trout (*Salmo aguabonita*), the game fishes in my district were introduced or exotic species.[1] Nothing was being done to assure the preservation of the native life forms included within my stewardship. No one even knew what they were! No one had really given the matter much thought. Virtually no attention was devoted to the study of the basic components of the biota. We were living in a make-believe world. The California Department of Fish and Game, as with most state fish and wildlife agencies of that era, was spending its resources painting the building while the foundation crumbled. Although my department as a whole seemed pleased with what was going on, I felt a strange foreboding and knew that, somehow, things had to change. During the summer of 1964 I returned to *A Sand*

County Almanac and reread "The Land Ethic" at leisure and in depth.

More than a decade of field experience gave Aldo Leopold's words new meaning. Within the principles which he so eloquently set forth I found a rational basis for approaching and solving the problems that perplexed and seemed so completely to overwhelm me. I felt I had within my grasp the basic components for making management programs address the entire biota, not simply the superficial popular demands which had so fully and frivolously consumed my time.

A Sand County Almanac clearly articulated my intimations that, to be really meaningful and to serve the long-term interests of the biota (and therefore the people), any management program worthy of the name must begin with the integrity of the land and water. Using this as a foundation, the resource manager can then build his pyramid upward, adding to this base the flora and fauna, and then build further to accommodate, finally, the species of economic and political interest. I had toyed with this possibility, but up to that point I had been unable to muster the courage necessary to buck the system. Leopold's grit, as well as his clear purpose and simple means, seemed to be the very thing I needed to gain this courage. I was especially motivated by his wry observation that "nonconformity is the highest evolutionary attainment of social animals."[2]

The universe as a whole is governed by the complex interaction of immutable and elegant physical laws. As I read through *A Sand County Almanac* it became evident that a set of natural laws, equally elegant and immutable, govern biological systems as well. The futility of trying to circumvent these laws for any appreciable length of time in an effort to achieve short-term economic and political goals became even more apparent to me.

It is important to remember that the early 1960s found the nation just entering the initial throes of a concerted conservation movement. The first (and very inadequate) Endangered Species Act was not passed until 1966, and the National Environmental Policy Act was several years away. Financial support for innovative programs did not exist, and administrative (and public) backing was almost totally lacking. The land ethic was not yet abuilding. But to me it seemed unconscionable for the department to be spending the great majority of its fisheries budget planting put-and-take trout and conducting meaningless creel censuses, while we had no clear understanding of the indigenous biotic community. When our pressing need to conduct biological surveys was brought to the attention of top administrators, even those with advanced degrees responded with either a blank look or a remark to the effect that the public and the legislature would never allow license money to be spent collecting "bugs or suckers," some of which we were in fact trying to eradicate in order to provide better angling.[3]

Adding insult to injury, the department had commissioned several fine wildlife scholars and administrators (headed by Starker Leopold) to prepare a plan to guide the department's activities until 1980. The principles of the plan were ecologically sound and, had they been followed, would have done much to reverse the status quo. However, politics and tradition often speak much louder than logic, and we stumbled on as if the plan had never been written.

An interagency meeting in Death Valley in April 1969 was called to discuss the status of the native fishes of the Death Valley hydrographic area (the Pleistocene drainages entering Death Valley originate in both California and Nevada). I learned there that my colleagues in the Nevada Department of Fish and Game (now the Department of Wildlife), National Park Service, Bureau of Land Management, and Fish and Wildlife Service faced a similar dilemma. Recognizing that at least a decade could be expected to elapse between recognition of a problem at the field level and the mustering of full financial and administrative support from government in an effort to solve it, we realized that we would have to move independently of bureaucracy. In effect, we moved underground. Had we waited for official support, it is absolutely certain that at least two, and probably five, fishes in two genera would now be extinct. We would have lost the last remaining individuals of the entire genus *Empetrichthys;* and as their habitats were destroyed, we no doubt would likewise have lost an unknown number of invertebrate species and plants.[4] And by "lost" I mean extinct—gone forever—in the words of Alfred Russel Wallace, "uncared for and unknown."[5]

A Pilgrim's Progress from Group A to Group B

The obvious futility of continuing to blindly repeat conventional management procedures caused me to look elsewhere for long-term solutions to problems that were growing worse with each passing day, and with no relief in sight.[6] Unfortunately, the biologically sound concept of conducting basic inventories of native species was hampered because the habitat integrity and species composition of the Owens and Walker river systems had already been significantly altered during the early part of the twentieth century. Even so, "The Land Ethic" could serve to establish the ideal that we should be striving to achieve: habitat integrity and an adequate and complete complement of native species. *A Sand County Almanac* literally charted our course: "A land ethic of course cannot prevent the alteration, management, and use of these 'resources,' but it does affirm their right to continued existence, and, at least in spots, their continued existence in a natural state" (204).

One could sense a change, although our management and research programs remained utilitarian. Actually, "economically or politically expedient" would better characterize them, since enlightened self-interest, the hallmark of utilitarianism, if carried to its logical limits, must lead to a policy of basic resource integrity and protection.

With the blessings of an understanding supervisor, the fish-hatchery system kept the wolves away from the door while I began to substitute species inventories for creel censuses. The initial survey revealed that three of the four fishes native to the Owens River system were either endangered or of indeterminate status.[7] Eventually, informal species recovery plans and similar nonconsumptive management programs were initiated, along with a plan designed to further self-sustaining wild trout. Major efforts were devoted to environmental protection. Additional efforts to "educate" the general public and to win the political support necessary for the perpetuation of an environmentally sound program of resource management have been undertaken. It is gratifying that this new overall approach appears to be successful and has not, as it was feared, provoked significant adverse public or political reaction.

Among the most profound of Leopold's observations is the following: "To promote perception is the only truly creative part of recreational engineering" and its corollary: "The only true development in American recreational resources is the development of the perceptive faculty in Americans. All of the other acts we grace by that name are, at best, attempts to retard or mask the process of dilution" (173, 174). It is the height of naivete to assume for even a moment that government can continue for long in a specific direction, irrespective of the righteousness of its cause, without the support of the people. The concept of public perception therefore becomes paramount.

Abraham Lincoln supposedly observed that understanding a problem constitutes half of its solution. I have come to believe, on the basis of my experience, that if *A Sand County Almanac* were read with an open and understanding mind by the staffs of the nation's fish and wildlife agencies (especially administrators), we would be well on our way as a nation to establishing a sound basis for dealing with virtually any of the resource problems which the future might bring. In effect, *A Sand County Almanac* would constitute the humanities component—the values and philosophy—of fish and wildlife management and provide a sound moral basis for drafting an ecologically responsible overall agency approach. Unfortunately, few of our colleges or universities provide undergraduate courses in environmental philosophy or ethics. Most agency biologists, if they gain an interest in the values and goals of resource management at all, do so through an insight essentially experiential in origin.

Although less so than in past years, it should come as no great surprise that Leopold's A-B cleavage remains alive and healthy within the fish and wildlife profession. "B" types, who regard the land as an integrated biota, mostly remain in academe, possibly because they are primarily research- and theory-oriented. "A" types, who regard the land primarily as a vehicle for producing a harvestable crop, abound in fish and wildlife agencies, because a love of hunting and/or fishing may have provided the major motivation for their entry into the profession, and their thinking very often has not progressed much beyond that.

The significance of this fact is often overlooked by analysts seeking to understand the history and evolution of the nation's

A Pilgrim's Progress from Group A to Group B

fish and wildlife programs. "A" types were quick to discover that the license buyers who funded most agency programs were also overwhelmingly of the A persuasion. It then became a simple matter to obtain the legislative, administrative, and financial backing consistent with economically popular programs. Thus, it is apparent why the very basic and ecologically essential non-game component of the biota was almost totally neglected in favor of economically valuable species.[8] To paraphrase Leopold here, we, in the agencies, fancy that game species support us, forgetting what supports game species (178). Obviously, this support is manifested in healthy land and a healthy, intact, over-all biota: soil, plants, invertebrates, nongame and game species alike. At this writing the A-B cleavage is still strongly repre-sented throughout fish and wildlife agencies. The A component is often given strong support by politically appointed commis-sioners, who then receive the accolades of the sportsmen's groups responsible for their appointments.

My transition from an A to a B philosophy in turn resulted in a major shift in my professional emphasis. In 1959, the Los An-geles Department of Water and Power applied to the California Division of Water Rights for a permit to divert several miles of lower Rock Creek—one of the eastern Sierra's most heavily used recreational streams, and an excellent fishery—into Crowley Lake in order to produce additional hydroelectric power from the Owens River Gorge power plants. Inasmuch as Los Angeles had already dried up several miles of the famous Gorge fishery during the early 1950s, the thought of losing additional stream mileage was unacceptable to me, especially since doing so would be in violation of new laws resulting from the Gorge project.

The very rough studies which we conducted in preparation for the court proceedings constituted one of the early predecessors of the Instream Flow Technologies in current use throughout the West. However, they were adequate to convince the court in what turned out to be the first water battle that the California Department of Fish and Game had ever won, and the first that the city of Los Angeles had ever lost.

My motivation then was almost entirely to preserve a fishery for an exotic trout species. I would work even harder now to accomplish the same thing, but primarily to preserve the basic stream ecosystem and its component flora and fauna. The brown trout is now part of this ecosystem, but by no means the most important part. I am consciously guided by Leopold's summary precept: "A thing is right when it tends to preserve the integrity, stability, and beauty of the biotic community. It is wrong when it tends otherwise" (224, 225).

Twenty years later, following the "energy crunch" of the 1970s, Congress passed legislation encouraging entrepreneurial development of small hydroelectric projects. This has placed in great jeopardy the entire stream system of the eastern Sierra Nevada. So far we have been successful in saving these habitats, and the motivation again derives from Leopold's maxim. This precept is directly applicable and effective in virtually any problematic situation that a resource manager may confront—whether to introduce an exotic species, or to build a road near a fragile habitat, or whatever.

In 1985, for example, the Federal Energy Regulatory Commission decided to conduct an assessment of potential small hydroelectric developments within the Owens River Basin. We knew from our own analyses that 88 percent of our stream mileage downstream from Forest Service wilderness boundaries had already been impacted by diversions of one type or another, and approximately 25 percent of the hydroscape no longer existed. To further disturb this tremendously valuable stream resource to produce a miniscule amount of electric energy seemed categorically unethical, as measured by Leopold's cardinal principle. And it should not require a study of any sort to prove it.[9] However, when I related this opinion to a project official, his response chilled me: "When you start talking about morality and ethics, you lose me." I find such a philosophy (or lack of it) guiding a high public official frightening, to say the least.

Another illustration of a practical application of Leopold's precept of biological integrity occurred in the late 1970s when the National Park Service examined its policies and decided that since trout are not indigenous to the lakes of the High Sierra, they would no longer be planted in park waters. The intent here was to be consistent with a basic park policy to preserve the in-

tegrity of natural biota, although most ecological damage resulting from trout planting had occurred years before when the first
trout introductions were made. The Department of Fish and
Game opposed the new park policy primarily because it constituted revocation of a historic state management procedure that
had never before been questioned. I sided with the Park Service
against my own agency because the Park Service policy was
clearly more in line with the fundamental directive of the land
ethic.

Later, we reached an ethically sound compromise. We are not
planting lakes in the national parks, but the Forest Service,
which administers High Sierra waters outside the park boundaries, has adopted an ecologically realistic and politically acceptable policy: continue to plant lakes that have been planted historically, but do not plant lakes hitherto barren of trout.

The concept of biological (environmental) integrity is comprehensive. As Leopold avers, "a land ethic changes the role of *Homo
sapiens* from conqueror of the land-community to plain member
and citizen of it. It implies respect for his fellow-members, and
also respect for the community as such" (204). Moreover, *Homo
sapiens* will benefit in the long run from a healthy biota. This
should provide enough of a utilitarian rationale to elicit political
support for ecologically sound decisions. We, too, are part of the
system, and if something is good for owls, trout, and lizards, it
will, in the final accounting, be good for us as well.

Aldo Leopold lamented the failure of the American public to
recognize the need for, and to implement, a land ethic. If he now
were to look down upon us, "perchance in some far pasture of
the Milky Way," it is likely that he would be gratified by what has
happened in the environmental field since he left us (101). The
environmental movement of the 1960s and 1970s was in no small
measure inspired by his philosophy. The National Environmental Policy Act, Endangered Species Act, Marine Mammal Protection Act, Clean Water Act, and similar laws passed by the states
were also in no small measure inspired by his philosophy. The
publication of this volume itself contributes to the growing
awareness of what he modestly called "conservation," and I predict that *A Sand County Almanac* will continue on well into the

future as one of the major forces shaping the nations's environmental policies.

At the state level, we see an increasing emphasis on nongame species. In California, for instance, a Natural Diversity Data Base, conducted in cooperation with The Nature Conservancy, is drafting an inventory of the state's flora and fauna. Although embryonic at this stage, and small in comparison with traditional programs, progress in this direction is heartening. A strong effort at top administrative levels to achieve funding sources other than hunting and fishing licenses, and earmarked for the management of nongame wildlife, reflects the idea that wildlife resources are an amenity for all the people, and not just the property of consumptive users. Additional efforts to acquire key habitats, and to coordinate interagency activities, are also underway.

It is a credit to Aldo Leopold that today increasing concern is shown for those creatures that share a common habitat with us. More and more scientists and lay persons alike conceptually integrate species with habitat in our efforts to preserve biological diversity.[10]

There are those who might debate the practicability of the land ethic for contemporary fish and wildlife management programs. I would urge them to remember that Aldo Leopold was a great naturalist and wildlife professional whose humility allowed him to probe an unspectacular local ecosystem, to be informed by what he found there, and to record it with a beauty and clarity of expression unadorned by needless sophistication. His wisdom was forged by a lifetime of field experience and reflection on his own mistakes, and by his careful husbandry of a small parcel of poor land.

A loathing of freeways and asphalt has kept me in the field for more than thirty-five years now. The solitude of sand, sage, and pine has allowed me to hear "a few notes" of that "vast pulsing harmony—its score inscribed on a thousand hills, its notes the lives and deaths of plants and animals, its rhythms spanning the seconds and the centuries" (149). And the lessons of the backward sand county farm have allowed me to approach and solve the management problems presented both by an ever-increasing and demanding public, and those who would degrade or destroy the habitat that is basic to the recreational resource.

Because I know that the land ethic works well in the field, I have gained much satisfaction in recent years by relating my management experiences to students in places ranging from Nuevo Leon in Mexico to British Columbia in Canada, encouraging them to avail themselves of the wisdom born within the soil, waters, and winds of the sand counties. Learning that a nucleus of environmental philosophers was similarly engaged took me to a conference at the University of Georgia in October 1984. Not long after, J. Baird Callicott invited me to lecture to resource management classes and to participate in Earth Week activities at the University of Wisconsin–Stevens Point. My honorarium consisted of a visit to "the shack." While dinner was being prepared there I stole away by myself and stood beside the "good oak" with Starker and his dad. There was really no reason to speak; I had long ago been well taught. Then through the clouds of the early evening I heard goose music as a single pair of honkers set their wings over the Crane Marsh and swung low over the Wisconsin River. Somehow I knew they would always do this, and I glanced for a moment into eternity. Like the "round river," I had come home.

Notes

1. Within the fisheries profession an introduced species is defined as one introduced from another drainage, whereas an exotic species is endemic to, and introduced from, another continent.

2. Aldo Leopold, *Round River* (New York: Oxford University Press, 1953), 8.

3. See Stephen R. Kellert, "Social and Perceptual Factors in Endangered Species Management," *Journal of Wildlife Management* 49 (1985): 528–36.

4. Edwin P. Pister, "Desert Fishes and Their Habitats," *Transactions of the American Fisheries Society* 103 (1974): 531–40; Robert Rush Miller and Edwin P. Pister, "Management of the Owens Pupfish, *Cyprinodon radiosus,* in Mono County, California," *Transactions of the American Fisheries Society* 100 (1971): 502–9.

5. Alfred Russel Wallace, "On the Physical Geology of the Malay Archipelago," *Journal of the Royal Geographical Society* (London) 33 (1863): 217–34. See also Holmes Rolston, III, "Duties to Endangered Species," *Bioscience* 35 (1985): 718–26.

6. Edwin P. Pister, "Desert Pupfishes: Reflections on Reality, Desirability, and Conscience," *Environmental Biology of Fishes* 12 (1985): 3–12.

7. Ibid.

8. Edwin P. Pister, "A Rationale for the Management of Nongame Fish and Wildlife," *Fisheries* 1 (1976): 11–14.

9. Edwin P. Pister, "Endangered Species: Costs and Benefits," *Environmental Ethics* 1 (1979): 341–53.

10. Jack E. Williams et al., "Endangered Aquatic Ecosystems in North American Deserts With a List of Vanishing Fishes of the Region," *Journal of the Arizona-Nevada Academy of Science* 20 (1985): 1–62.

11 * The Legacy of Aldo Leopold

WALLACE STEGNER

In 1949 a University of Wisconsin professor of ecology named Aldo Leopold published a book of essays called *A Sand County Almanac*. It is a famous, almost holy book in conservation circles. Conservationists know Leopold also as the originator (with Bob Marshall) of the idea of permanent wilderness reservations in the United States, and as the father of Starker and Luna Leopold, environmentalists nearly as distinguished as himself. But *A Sand County Almanac* is the best way to know him. When this forming civilization assembles its Bible, its record of the physical and spiritual pilgrimage of the American people, the account of its stewardship in the Land of Canaan, *A Sand County Almanac* will belong in it, one of the prophetic books, the utterance of an American Isaiah.

"To understand the fashion of any life," Mary Austin said in *The Land of Little Rain*, "one must know the land it is lived in and the procession of the year." [1] Leopold's essays follow with love the procession of the year in a Wisconsin county raped and abandoned by the loggers, and slowly recovering. They cry woe for what we have done to the earth's fairest continent and to the plant and animal species we found in it. They also analyze the errors of understanding and attitude and aim that have led us to failure just short of calamitous; and they suggest a change of heart and mind, a personal conversion, a reversal of individual

This essay originally appeared in *Wilderness* (Spring 1985), under the title, "Living on Our Principal."

233

and communal carelessness, which might lead to changes in public policy and let us still salvage and partway reclaim our earth and ourselves. What Leopold hoped for was the gradual spread of a "land ethic," built upon scientific understanding of the earth and earth processes, but infused with a personal, almost religious respect for life and for the earth that generates and supports it.

On the face of it, a land ethic does not seem much of a defense against the greed that has consistently moved us; nor against the conception of land as a commodity to be bought, sold, and exploited without regard for its living continuities; nor against the twentieth-century technology that gives us such power to destroy. Nevertheless, the land ethic goes to the heart of the matter. Leopold was not one of those throbbing Nature lovers who, as he said, "write bad verse on birchbark" (167). He was a scientist, one of the first to profess the new science of ecology. As a forester and game manager he had seen for himself, in many places, the slow death of the land, and he knew that our unchecked effort to make everything safe and comfortable for our own species at the expense of all others could eventually destroy us along with the earth we depend on.

He told us many things that have by now become truisms of the environmental movement. "The old prairie," he wrote, "(or any biome), lived by the diversity of its plants and animals, all of which were useful because the sum total of their co-operations and competitions achieved continuity" (107). He had the faith in the natural order that some economists have in the free market. Human interference and management break down the food chains and tear the web of intricate interdependence. Unless done with care, discretion, and restraint, they are sure to result in land-sickness, and can result in total land death, as in the man-made deserts of the world.

Not only the burn-plant-and-move-on agriculture of some primitive peoples, but our own scientific agriculture is capable of great long-range harm. Monoculture, in which all the roots are at the same soil level and all the plants are subject to the same pests; over-grazing; manipulations of surface and subsurface water; the cutting of timber on the watersheds; even deep plow-

The Legacy of Aldo Leopold

ing—these can bring on, severally or together, soil exhaustion, erosion, salinization, depletion of the water table, the death of species, and the spread of liberated or imported weeds.

We kill the thing we love, Leopold said, "and so we the pioneers have killed our wilderness. . . . The Congressmen who voted money to clear the ranges of bears were the sons of pioneers. They acclaimed the superior virtues of the frontiersman, but they strove with might and main to make an end of the frontier" (137).

None of that sounds new in 1985. Except in its scientific justifications, it was not new even in 1949. The lament for our wastefulness and greed runs like a leitmotif through our literature from very early times. James Fenimore Cooper, our first significant novelist, voiced it before our frontier had moved beyond the Great Lakes. Read the scene in *The Pioneers* (1823) in which the townspeople of an upstate New York settlement, in their frenzy to reap God's plenty, fire into flocks of passenger pigeons with artillery. Read the passage in *The Prairie* (1827) in which Cooper asks passionately what the axemen will do when they have cut their way from sea to sea. (We know what some of them do. They sit in Eureka bars and foam at the mouth about Redwoods National Park, or they slip into the Avenue of the Giants after dark and with their chainsaws, out of sheer spite, girdle thousand-year-old trees.)

Nature, and the love of it, is a major theme in others than Cooper. It is there in Thoreau, Emerson, Muir, Burroughs, Whitman, and it continues into the present as an unbroken double song of love and lamentation. Lamentation is the conscience speaking. Below the energy, enthusiasm, and optimism of our early years—below the Whitman strain—there is always the dark bass of warning, nostalgia, loss, and somewhat bewildered guilt.

So it was not for its novelty that people responded to Leopold's call for a land ethic. It was for his assurance, an assurance forecast much earlier in the work of George Perkins Marsh and John Wesley Powell, that science corroborates our concern, not our optimism: that preserving the natural world we love has totally practical and unsentimental justifications. Many Americans will discount poets and mistrust barefoot Nature lovers, and schoolmarms who cry "Woodman, spare that tree!", and Indians who revere the earth as their mother; but they find it harder to shrug off what they think of as the "practical" evidence of scientists.

It was Leopold's distinction that he combined and reconciled two strains of thought. He had no romantic revulsion against plowing, or cutting trees, or hunting birds and animals, or any of the things we do to make our living from the earth. He was only against the furious excess of our exploitation, our passion to live on our principal. It struck him as sad that unexploited land, land left alone, seemed actively to offend our continent-busters. (We hear them still. They are the people who resent what they call the "locking-up" of resources.)

"Progress," Leopold wrote, "cannot abide that farmland and marshland, wild and tame, exist in mutual toleration and harmony," and so we have assiduously drained most of the marshes where ducks, geese, and cranes used to add their seasonal magic to our life (162). Plowmen and stockmen were never able to resist putting every square rod of grassland to use and overuse, with the result that many once-luxuriant prairie plants are now extinct, and others survive only along fencelines where the plow could not reach, or within railroad rights-of-way fenced against both plows and grazing animals. The prairie biome is weakened in ways we can only guess.

Unlike many early conservationists, including John Muir, Leopold was as suspicious of recreation as he was of logging and agriculture. He noted that the impulse to refresh ourselves out of doors has damaged many of our remaining wild places—and he wrote before off-road vehicles had carried damage close to ruin. The lure of tourist money, he said, is a gun pointed at the heart of the wild, and he was right.

One example: for the last thirty years we have had to watch the congressional delegation of Utah, one of the most marvelously endowed of the fifty states, do its best to cheapen and imperil the magnificence of southern Utah by promoting freeways on which tourists could view it at seventy miles an hour. The only really creative part of recreational engineering, Leopold

The Legacy of Aldo Leopold

insisted, is to promote *perception*. "Recreational development is a job not of building roads into lovely country, but of building receptivity into the still unlovely human mind" (176–77).

With Thoreau, he believed that "in wildness is the preservation of the world," and he believed in wildness not because it is sublime but because in it is all our hope, the reservoir of all future possibility.[2] "Wilderness was never a homogeneous raw material. It was very diverse, and the resulting artifacts are very diverse. These differences in the end-product are known as cultures. The rich diversity of the world's cultures reflects a corresponding diversity in the wilds that gave them birth" (188).

He was one who made us begin to understand that wilderness is indispensable for science and survival. We need to keep the wild gene pool diverse and healthy, if only because our domesticated species are subject to devastating pests, and may need some genetic assistance.

We need to understand, better than we do, why some species have disappeared forever and some have proliferated as weeds. Without rejecting Emerson's definition of a weed as a species for which man has not yet found a use, Leopold might have pointed to weeds as species that, imported into environments where they have no competitors, or liberated from their natural competitors by our intervention, have gone out of control. To counter weeds, as well as deal with the other ills we bring on, we resort to what he called "land doctoring." "The art of land doctoring is being practiced with vigor, but the science of land health is yet to be born" (196).

New or old, those ideas were heretical in 1949, and in some quarters still are. They smack of socialism and the public good. They impose limits and restraints. They are anti-Progress. They dampen American initiative. They fly in the face of the faith that land is a commodity, the very foundation-stone of American opportunity, and that it must be put to the use that the assessor's office would certify as the highest use—meaning economic use. Oil drillers, miners, lumbermen, stockmen (all deeply involved in the rape of the Public Domain), and recreationists, just as deeply involved, and farmers concerned to get maximum short-term productivity out of their family or corporate farms, and water users busily pumping down the water table of desert valleys, and developers hungry to squeeze the ultimate building site out of a parcel, have not adopted Leopold's views with enthusiasm.

Nevertheless, he thought there might be an outside chance of correcting our course—if we can expand our ethics to take in land as well as people. Those twelve slave girls whom Odysseus hanged for their indiscretions roused no sympathy or guilt in either Odysseus or Homer. We have developed a tenderer conscience about such matters in three thousand years, we inherit the worldwide revulsion against slavery that took place in the nineteenth century, we have been taught to respect human rights no matter whose. If we can do as much about understanding the rights of the natural environment we have a chance.

That was the way Aldo Leopold saw it at mid-century. His book has taught tens of thousands of Americans to see it in the same way, and the nearly forty years since the publication of *A Sand County Almanac* have surely been years of increasingly acute environmental awareness. But as the century explodes toward its end we may well ask if the strain of thought that Leopold developed and stimulated has had much effect on the vague, changeful, and uninformed entity we call the public mind.

Are we any closer than we were in 1949 to a land ethic as widely acknowledged (however widely evaded) as our commitment to civil rights? Or have we, indulging habits learned on our careless frontiers and given continuing force by the Reagan administration, pushed closer to the showdown that, if not so dramatic and totally annihilating as the atomic showdown we simultaneously risk, will surely bring a sharp decline in the quality of American life?

Certain inescapable facts are relevant. At mid-century the population of the United States was 150 million. Since then we have added more people than the whole country contained in 1900, and now number above 230 million. That 53 percent growth above the 1950 base has meant what we in our lives have witnessed: more mouths to feed; a greater and greater demand for energy to run our industries and our homes; more digging for minerals and coal, more cutting of timber, more drilling for oil

on and off-shore; more automobiles to carry us around and foul
the air; more paving-over of land for freeways, roads, driveways,
parking areas, and tennis courts; more orchards and fields gone
into subdivisions or industrial plants or shopping centers; more
bombing and missile ranges in the deserts; more ski resorts in
the mountains; more smokestacks and acid rain; more Love Ca-
nals and toxic waste sites and more dirty politics to cover them
up; less green, less space, less freedom, less health; more intensive working of the soils still left in agriculture; a longer and
longer stretching of a rubber band not indefinitely stretchable; a
temporarily more comfortable, ultimately less plentiful, increasingly less spiritually rewarding life.

Precisely because we could not ignore the symptoms, we managed some corrections, especially during the 1960s and 1970s,
before the Reagan administration undertook to dismantle the
gains. Most of the corrections we measure by federal laws; and
though Leopold himself warned us against expecting the government to exercise our conscience for us, we would be in infinitely worse shape without the Clean Air Act, the Clean Water
Act, the Strip Mining Act, the Wilderness Act, the Federal Land
Protection and Management Act (FLPMA), the Land and Water
Conservation Fund Act, the Alaska Native Lands Act, and many
other laws that have been enacted since Leopold wrote. Without
those laws to evade or circumvent or confront, James Watt and
Anne Gorsuch Burford would have done even greater harm than
they succeeded in doing before being forced out of office.

Also, in the very fact that they were forced out of office, that
they outraged the environmental conscience of enough people
to do themselves in, there is evidence that Leopold's hope for a
growing individual and public awareness was not totally naive.
Their fall at least enforced caution on the administration. It discovered that it couldn't ride roughshod over the body of environmental law. One hardly expects a change in policy, but one
expects greater circumspection.

Other indications: when Leopold wrote, there were only a
few principal environmental organizations, each devoted to a
single issue and none (not even the Audubon Society, which had
begun that way) politically militant. In addition to the Au-

dubon, which had been organized to protect birds from plume hunters, there was the Sierra Club, for mountaineers; the Izaak Walton League, for fishermen; the Wilderness Society, devoted to wilderness and wildlife; and the Save the Redwoods League, devoted to just what its title indicated.

Now there are dozens: the Nature Conservancy, the Trust for Public Land, the Open Space Trust, the Environmental Defense Fund, the Natural Resources Defense Council, the League of Conservation Voters, the Sempervirens Fund, Trout Unlimited, Ducks Unlimited, the National Wildlife Federation—a long list, plus a thousand regional and local defense committees organized to protect a valued local resource from destructive exploitation.

The spirit of all those groups is aroused and aggressive. James Watt himself noted that one of the first effects of his application of Reagan principles was to double or triple the membership of the conservation organizations—the "environmental extremists"—and to galvanize their political activism. Watt and Gorsuch walked into a buzzsaw that they had apparently thought was a wreath of wildflowers.

Thus, since Leopold's day we have seen not only a worsening of all the environmental sicknesses, but the development of an environmental constituency that grows steadily larger and more militant. Our national parks have always had a loyal party of backers, in and out of Congress, and until 1981 (with brief lapses during the Truman and Eisenhower administrations) they were steadily added to to provide for a growing public need. Essentially the same constituency pushed through the Alaska Native Lands Act, which, though signed by Reagan, was really Carter's last Act. The same constituency had brought about, in 1964, the wilderness system that Leopold had first proposed for the Gila country of New Mexico, back in the 1930s.

That old party of loyalists has been immensely augmented in the past twenty years as people began to protest dirty air, poisoned water, and creeping uglification. But how much evidence is there that a real conversion has taken place in a substantial part of the public? Earth Day fired up the college campuses briefly and subsided to something less than a conflagration. Is there

The Legacy of Aldo Leopold

anything to the environmental movement now besides the activity of those irreconcilable "environmental extremists" who so exasperated and frustrated Mr. Watt? Is environmentalism any more than the effort of a comfortable middle class to preserve its amenities? Do workers share in it? Blacks? Chicanos? College students? The young? The old? Is there perhaps even a current backlash against environmentalism?

I can ask those questions but not answer them. I am not a statistical man, and have no numbers, and probably would not trust them if I had them. But I have to believe that we have done better in pushing for and enacting environmental legislation than we have in creating and spreading a real environmental conscience. We have reached the immediately educable. The ineducable, who I am afraid include Mr. Reagan, continue to live by the myths of old-fashioned American enterprise, and are either indifferent or actively hostile.

The environmental lobby, potent during the sixties and seventies, and still formidable, perhaps does represent the views of a special-interest minority rather than a broad and united public opinion. That it has been a disinterested special-interest group, bent upon securing a truly public good and resisting special privilege, gives it a suspiciously pinkish tinge to those who cannot conceive of disinterested action. It is hard to resist publicly because it asks nothing for itself, but it is easy to disparage as an impractical and probably socialist frill.

In our time we have seen the environmentalism that used to be a bipartisan movement fall by default to the Democrats. No Republican administration in recent memory has been better than tepid on conservation issues, and most have been reluctant or opposed. None has picked up the mantle of Theodore Roosevelt—and even Roosevelt came under fire from many in his own party who resented what "that damned cowboy" was doing to the country.

Herbert Hoover and his Secretary of the Interior Ray Lyman Wilbur did their best to give away the Public Domain to the states. Warren G. Harding and *his* Secretary of the Interior Albert B. Fall gave us the Teapot Dome scandal. Dwight D. Eisenhower and *his* secretary of the Interior Douglas McKay threw the public lands to the corporate wolves. Ronald Reagan and *his* Secretary of the Interior James Watt sponsored the blind, greedy, and reactionary outburst from western stockmen that was known as the Sagebrush Rebellion, and did their best to open up the public lands on- and off-shore to coal and oil leasing at bargain basement prices. And even in Democratic administrations, but especially in Republican ones, the U.S. Forest Service for which Aldo Leopold worked many years of his life has persisted in its policy of board feet, at whatever cost to the environmental values that Leopold helped to define.

But if Republicans have not, as a party, espoused conservation, many individuals who would like to be Republicans have done so, and the Democratic Party has happily adopted the environment as its own to save. The trouble is—and this is a commentary on how thinly distributed the environmental conscience is—it has not been a winning political issue. In the ordinary voter's hierarchy of problems it pretty clearly ranks below the economy, the arms race, the nuclear threat, the women's movement, social security, civil rights, and other matters.

Because environmental degradation is, as Stewart Udall said, a quiet crisis, a creeping calamity, it makes news only when it crests in some dramatic episode such as the Love Canal or Rita LaVelle's handling of toxic waste sites. It makes news because of people's fear or someone's misfeasance, not because of people's consciences.

And fear is much more likely to be aroused by poisoned air or polluted water than by diminishing soil fertility or increased erosion. The single environmental issue significant in the 1984 political campaigns was acid rain, and most Americans outside the affected Northeast probably didn't have much feeling even for that. Clean up the worst of the waste sites and improve the unbreathable air and the polluted groundwater, and the public environmental conscience might all but evaporate. Back in Ohio and Indiana where the sulphur dioxide comes from and where cleaning up the stack emissions might cost every ratepayer some money, it has already all but evaporated.

What is more, the air, and to a lesser extent the water, exist in our minds as free public necessities. Land is something else:

The Legacy of Aldo Leopold

property. A land ethic calls not only for corporate scrupulousness but for a modification of our acquisitive impulses, something we have not yet learned with regard to real estate dealings. Your saved open space and rescued air is my spoiled opportunity, and this is the land of opportunity, isn't it? And opportunity is measured in dollars, isn't it?

Set up a new national park, with its opportunities for health, quiet, beauty, and the recreation of the spirit, against an opportunity for everyone to make a sum of hard dollars, and you know which opportunity most Americans will choose. Most people anywhere. Americans are no worse than other people. They have just had more practice in making money and wasting land, and more encouragement to go on doing so.

One area where a substantial number of people have had their minds changed is the preservation of wildlife and wildlife habitat. Not only the National Wildlife Federation, the largest of the environmental organizations, but hunters and gun clubs have realized the necessity of conservation if shooters are to have anything to shoot. And over the past thirty years there has been a flood of wildlife films that, being among the few things on television suitable for children's viewing, have beyond question educated many young people to feel kindly toward wildlife, including the predators with whose evil nature parents used to scare children a generation ago.

These films appeal to the Bambi impulse. They go in for cute pairs of bear cubs, river otter pups, or cheetah kittens who go out into their wilderness environment and through comic misadventures learn about life. But better sentimentality than ruthlessness, better an impulse to pat and cuddle than to scream and shoot. By making a niche for itself in television programming, wildlife may have done more to save itself than all the devoted organizations together.

But cute otter pups, or a spotted fawn curled up in the concealing grass, threaten no one. A national park or a wilderness reservation that blocks development may seem to. The closer we are to economic insecurity, the more willing we are, by and large, to sacrifice grasslands for the jobs and short-term profits of a strip mine. The harder our noses are to the grindstone, the more willing we are to suffer the ugliness and poisons that emanate from the smokestack plant on which we depend for a paycheck. The lunch-bucket argument has always been the most effective argument, even when it is misapplied, against environmental restraints.

Thus some citizens of Moab, Utah, and Hanford, Washington, vie for the location of a nuclear waste dump in their community. Thus workers in Ohio and Indiana resent being told that their jobs are producing acid rain. The attitudes of bosses need not be considered, for they do not have to live with what they create. But an environmental conscience in people who live close to the margin may often seem more self-sacrificial than anyone has a right to ask.

Not only the struggling and desperate are hard to reach and persuade. Those who use the deserts, mountains, and beaches for wheeled recreation are at least as destructive and less forgivable. When we wonder whether the barbarians who tear up fragile desert ecosystems with their frenzied trail bike rallies ever stop to think what they are doing, we remember the remark of a nineteenth-century politician in another context. "Does a dog stop for a marriage license?"

We are a long way yet from where Leopold hoped we might one day arrive. Corporate greed may be controlled by legislation—though it is not likely during the present administration. Special lands may be set aside, as in the past, as national parks, wildlife refuges, seashores, lakeshores. But we are running out of the raw material for such reservations, and there is no more where it came from. Moreover, the tighter the pinch, the more persuasive the lunch-bucket argument against "unproductive" stewardship. The only real success, the creation and spread of an environmental conscience which will be as angered by the abuse of land as most consciences would now be angered by the abuse of a child—that may have to wait a few hundred or a few thousand years, and we may not have that much time.

The land ethic is not a widespread public conviction. If it were, the Reagan administration would not have been given a second term. The lunch-bucket argument, and the deification of greed as the American Way, reelected by a landslide one of the

The Legacy of Aldo Leopold

most antienvironmental presidents in our history. Aldo Leopold would not be encouraged.

There will always be those who believe that the Public Domain is owned and maintained by the rest of us for their exclusive exploitation and use. There will always be those who never breathed deeply of the wind off clean grassland, or enjoyed the shade of a hillside oak on a July day, or watched with pleasure as some wild thing slid out of sight, and who will never see the reasons for living with the earth instead of against it.

The number of functional illiterates that our free public education produces does not make us sanguine about educating a majority of the public to respect for the earth, a harder form of literacy. Leopold's land ethic is not a fact but a task. Like old age, it is nothing to be overly optimistic about. But consider the alternative.

Notes

1. Mary Austin, *The Land of Little Rain* (Boston: Houghton Mifflin, 1903), 88.
2. Henry David Thoreau, *The Writings of Henry David Thoreau,* vol. 9 (Boston: Riverside, 1983), 275.

12 ∗ Duties to Ecosystems

HOLMES ROLSTON, III

"A thing is right when it tends to preserve the integrity, stability, and beauty of the biotic community. It is wrong when it tends otherwise" (224–25).

"That land is a community is the basic concept of ecology, but that land is to be loved and respected is an extension of ethics" (viii–ix).

"The plant formation is an organic unit . . . a complex organism."[1] So Frederic Clements, a founder of ecology, concluded from his studies in the Nebraska grasslands. Henry Gleason, a botanist of equal rank, protested, "Far from being an organism, an association is merely the fortuitous juxtaposition of plants."[2] Leopold takes a middle route between these extremes.[3] The ecosystem is a "biotic community." Moreover, moving from what *is* the case to what *ought* to be, Leopold argues a land ethic, duties toward ecosystems.[4]

Clements' description has seemed implausible to most ecologists, but if correct, duties to a superorganism could plausibly follow, since ethics has classically felt some respect for organismic lives. Gleason's description has seemed as simplistic as Clements' is overdone, but if correct, duties to ecosystems would vanish. There can be no obligations to a fortuitous juxtaposition. What of Leopold's biotic community? Most ecologists have coordinated thought that an ecosystem was a real natural unit, a level of organization above its member organisms.[5] Is the description plausible? Do prescriptions follow?

John Passmore, a philosopher entering the argument, thinks that only paradigmatic human communities generate obligations. "Ecologically, no doubt, men form a community with plants, animals, soil, in the sense that a particular life-cycle will involve all four of them. But if it is essential to a community that the members of it have common interests and recognize mutual obligations, then men, plants, animals, and soil do *not* form a community. Bacteria and men do not recognize mutual obligations, nor do they have common interests. In the only sense in which belonging to a community generates ethical obligations, they do not belong to the same community."[6] Passmore is assuming that the members of a morally bound community must recognize reciprocal obligations. If the *only* communal belonging that generates obligations is this social sense, involving mutual recognition of interests, then the human community is the sole matrix of morality, and the case is closed. We owe nothing to nonhumans, much less to ecosystems. But Leopold wants to open the question Passmore thinks closed. Extending the logic of ethics beyond culture, can mutually recognized obligations and interests be replaced by respect and love for ecosystemic integrity, stability, and beauty? Can a community per se count morally?

A first consideration is that the organism is a model of cooperation while the alleged ecosystemic "community" seems a jungle where the fittest survive. Fully functioning persons can be expected to cooperate in the deliberate sense, and interhuman ethics has admired being kind, doing as you would have others do to you, mutually recognizing rights, or calculating the greatest good for the greatest number. But symmetrical reciprocity drops out when moral agents encounter amoral plants and animals. To look for considerate cooperation in the biotic community is a category mistake, expecting there what is only a characteristic of culture.

But at least nondeliberate *cooperation* is an admirable feature in organisms, who seem to command ethical respect because, from the skin in, they are models of coaction. The heart cooperates with the liver, muscles with the brain, leaves with the cambium, mitochondria with the nucleus. Life is contained within individualized organisms, notwithstanding colonial species, slime molds, and other minor exceptions. Respect for life, therefore, ought to attach to individuals, where the "integrity, stability, and beauty" is.

From the skin out, everything is different. Interactions between individuals are nothing but *struggle*. Each is out for itself, pitted against others. Carnivores kill herbivores, who consume the grasses and forbs. Young pines smother each other out. Black walnuts and *Salvia* shrubs secrete allelopathic agents that poison other plants. Apparent harmony in ecosystems is superficial. Chokecherries benefit redwing blackbirds, but the fruits are bait and a gamble in reproductive struggle. Neither plants nor animals are moral agents, and to regard carnivore or *Salvia* behavior as "selfishness" is a mistake, just as much as to expect deliberate cooperation. We are not faulting organisms. Still, the organism is parts admirably integrated into a whole. The ecosystem is pulling and hauling between rivals, no admirable community. To adapt Garrett Hardin's phrase, there is tragedy on the commons.

Such a picture accentuates the skin-in cooperation and the skin-out conflict. Ecology refocuses both the description and the resulting prescription. The requirement that parts in wholes "help each other out"—charitably in culture and functionally in nature—makes another category mistake, trying to assimilate to civilization what needs to be admired in wild nature. Were cooperation the criterion, we would admire the elephant's heart and liver, and even admire the integrated whole elephant, only to despise the elephant's behavior, since the integrated elephant-unit consumes all the bamboo and acacia it can and tramples the rest of the gallery forest in majestic indifference. The elephant is coordinated within to struggle without, and would-be admirers are left ambivalent about individuals. The individual is an aggrandizing unit as much as a cooperative one. Such units propel the ecosystem.

The deeper problem seems to lie in the axiom that everything is pushing to maximize itself, with no further determining forces. In fact, although aggrandizing units propel the ecosystem, the system limits such behavior; there is a sufficient but contained place for all the members. Imposed on organisms from the upper organizational level (if indeed each species increases until "stopped") this containment can seem more admirable than the aggressive individual units. The system forces what cooperation there is, embedding every individual deeply in coaction.

What we want to admire in nature, whether in individuals or ecosystems, are the vital productive processes, not cooperation as against conflict, and ethicists will go astray if they require in nature precursors or analogs of what later proves admirable in culture. We want to value the lush life ecosystems maintain, and the question of "helping each other out" is at most going to be a subset of this more significant issue, if "helping" is an appropriate category at all.

Painting a new picture on the conflict side, even before the rise of ecology, biologists concluded that to portray a gladiatorial survival of the fittest was a distorted account; biologists prefer a model of the better-adapted. Although conflict is part of the picture, the organism has a situated environmental fitness, including many characteristics that are not competitive for resources or detrimental to neighbors, as when some elephants survive heat or drought better than others. The elephant fits the savannas just as much as its heart fits its liver; there is equal fittedness within and without. There are differences: the heart and the liver are close-coupled; remove either and the elephant dies. Elephants and savannas are immediately weak-coupled. Elephants migrate from savannas to forests, or can be removed to zoos. But *Loxodonta africana,* removed from the selection pressures where the species evolved and its vigor is retained, soon dies too. Savannas and forests are as necessary to elephants as hearts and livers. The more satisfactory picture is of elephants pushing to fit into a system that provides and imposes sufficient containment.

The environmental necessity involves conflict, selection pressure, niche-fittedness, environmental support; the organic necessity involves cooperation, functional efficiency, metabolically integrated parts. The skin-in processes could never have evolved, nor can they remain what they are, apart from the skin-out processes. Elephants are *what* they are because they are *where* they are. What we mean by a *community* as a different systemic level from an *organism* includes these weaker, though not less valued or fertile, couplings. The two levels are equally essential. Adaptedness covers both. This invites respect for the ecosystemic processes quite as much as organismic processes. There seems no reason to admire the inside and depreciate the outside. Else we are including only half the truth about life. In result we will mislocate our sense of duty.

Even in human society conflict is not always taken as evil. Every academic pledges to keep the critical process open, wishing the "attacks" of those who can constructively (if first destructively) spot his or her flaws. Like business, politics, and sports, ecosystems thrive on competition. In a natural community the cougars are the critics (if we may put it so) that catch the flawed deer, and thereby build better ones, as well as gain a meal. Alternatively, the fleet-footed deer test out any cougars slow enough to starve. There is violence in the one process, and the other ought to be civil. Ideas die in the one realm, while individuals die in the other. Justice and charity may be relevant in culture and not in nature. But in both communities, helping is subtly entwined with competition. There is a biological, though not a cultural, sense in which deer and cougar cooperate, and the integrity, beauty, and stability of each is bound up with their coactions. Ecosystems are not of disvalue because contending forces are in dynamic process there, any more than cultures are.

Predator and prey or parasite and host require a coevolution where both flourish, since the health of the predator or parasite is locked into the continuing existence, even the welfare, of the prey and host. The one must gain maximum benefit with minimum disturbance of the other; it is to the advantage of predator and parasite to disturb prey and host species minimally. Although individuals are weakened or destroyed, if the disturbance is too great, the prey will evolve to throw off the predator, the host the parasite, or they will become rare, or extinct, to the disadvantage of predator and parasite. Parasitism tends to evolve into mutualism (e.g., cellulose-digesting bacteria in the ungulate rumen, algae and fungi synthesizing lichens, or *Chlorella* alga within the green hydra, *Hydra viridis*).

It seems doubtful that "plant defenses" are that and nothing more. Plants regulate but do not eliminate the insects that have

Duties to Ecosystems

coevolved with them. Pollinators and fruit eaters yield benefits to and benefit from the plants they serve; insect consumers eat less than 10 percent of the terrestrial biomass upon which they graze (certain outbreaks excepted), and insects (even outbreaks of them) seem often to provide benefits of which we are as yet little aware, as we once were unaware of the benefits of fire. Aphids secrete sugars that stimulate nitrogen-fixing bacteria in the soil, and short-lived insect grazers permit to long-lived plants rapid nutrient recycling, something like that accomplished more slowly by seasonal leaf-fall and decay. Some species of grasses coevolved with grazing ungulates; neither can flourish (or even survive) without the other.[7] Here too, as with predators and prey, being eaten is not always a bad thing. Selection pressures will routinely drive adaptation and counteradaptation toward minimum disturbance, that is, to check competitions by forced cooperations.

An ecosystem is an imposing critical system, with a dialectic that keeps selection pressures high, enriches situated fitness, evolves congruent kinds in their places with sufficient containment. The ecologist finds that ecosystems objectively are satisfying environments (organismic needs are not all gratified, but enough are for species long to survive), and the critical ethicist finds (in a subjective judgment matching the objective process) such ecosystems to be imposing and satisfactory communities to which to attach duty.

It may be objected that an organism is a highly centered system; in contrast an ecosystem has no centeredness at all. The one is a marvel; the other is a muddle. The "inside" coactions routinely look teleologically constructed. Before Darwin, that fooled people into believing in design, and, whatever we think now of creation, when we come to judge the present results, humans ought to value organisms as negentropic evolutionary achievements that simulate intelligent purpose. The cooperation praised earlier can be put more precisely: *integrated cybernetic autonomy* ought to be respected.

The struggle disliked earlier can be put more simply: ecosystems are primarily *stochastic process*. A seashore, a tundra, is a loose collection of externally related parts. Even after biologists soften conflict with adaptive fitness, a forest is mostly a game played with loaded dice. With measurable probability, red maple trees replace gray birch, and beech replace maples; each on average out-competes the other in the deepening shade. In a network of invasions there is minimal integrated process. The fox, its heart, and liver together need meat and water. But the members of a biotic community have no shared needs; there is only shoving. Or there is indifference and haphazard juxtaposition.

Much of the environment is not organic at all (rain, groundwater, rocks, nonbiotic soil particles, air). Some is dead and decaying debris (fallen trees, scat, humus). These things have no organized needs at all. The biotic sector runs by need-driven individuals interacting with other such individuals and with the abiotic and exbiotic materials and forces. Everywhere the system is full of "noise." The mathematics becomes complex; often the interactions are too messy to find regularities at all. Still, the issues are those of the distribution and abundance of organisms, how they get dispersed here and not there, birthrates and deathrates, population densities, moisture regimes, parasitism and predation, checks and balances. There is really not enough centered process to call common unity, which is why ecology has so few paradigms, and why duty directed here seems misplaced. There is only a catch-as-catch-can scrimmage for nutrients and energy.

The parts (foxes, wolves, sedges) are more complex than the wholes (forests, grasslands). Individual organisms are not decomposable; their parts (livers, hearts, culms, roots) crumple into waste outside their wholes. They cannot be divided without death, this attests to their heightened individuality. But in an ecosystem the parts (so-called) are transients. A vixen can move her den from forest to grassland and switch prey. Migrating birds inhabit no one community but range over dozens. Ecosystems are a continuum of variation, a patchy mosaic with fuzzy edges. Some interactions are persistent, others occasional; some drive coevolution; some do not. Species in any particular ecosystem do not have the same limits to their geographical distribution; their tolerance regimes differ. There are few obligate associations. A *Juncus* species can suffer a blight and its "place" be taken by a *Carex*. An ecosystem is often transitional and unstructured;

that makes it (some say) doubtfully a natural kind at all. So far from being a satisfactory community, an ecosystem is rather sloppy.

From this perspective, the units counted as ecosystemic "parts" have more integrity than the system in which they reside. To attach duty to the loose-coupled system would misplace duty, like valuing a social institution (a business firm, a state legislature) or even a casual collection (college alumni on tour) more than the individual persons who constitute these groups. The centers of autonomy, meaningful response, satisfaction, intrinsic value, lie in the persons. However much society supplies a context of support and identity, it is *egos* that count in human ethics. By parity of reasoning, though there are no egos in nonhuman nature, we should count the nearest thing: *selves,* somatic if not psychological selves. The focal point for cultural value is the high point of individuality: the *person.* The moral focus in ecosystems should be the high point of integrated complexity: the *organism.*

This too is a picture that ecological science refocuses. Admiring concentrated unity and stumbling over environmental looseness is like valuing mountains and despising valleys. Unity is admirable in the organism, but the requisite matrix of its generation is the open, plural ecology. Internal complexity arises to deal with a complex, tricky environment. Had there been either simplicity or lock-step concentrated unity in the surroundings, no creative unity could have been composed internally. There would have been less elegance in life.

Rapid and diverse insect speciation, resulting in highly specialized forms, is a response to increasing niches in varied topographies. Complex plant biochemistries arise to produce and to offset allelopathic agents, increasing heterogeneity. The primate brain, integrated with hands and legs, is a survival tool in a "jungle." Using instinct and conditioned behavior lemurs "figure out" probabilities, there is that much order, and contingency enough to churn the evolution of skills.[8] The environment is not capricious, but neither is it regular enough to relax in. One always needs better detectors and strategies. When lions are caged, their brains degenerate within a decade; one has taken away their jungle. Simple, little-changing environments usually result in stagnation across millennia. Dialectic with the loose environment (rich in opportunity, demanding in know-how) invites and requires creativity. The individual and the environment seem like opposites; they are really apposites; the individual is set opposed to its world but is also appropriate to it.

Further, a lack of centeredness or sharp edges does not mean a lack of relational complexity. Leopold is right to insist, against Gleason's fortuitous juxtaposition, that "the individual is a member of a community of interdependent parts" (203). Ecosystems are not as coherent as organisms, but not randomly fortuitous either; they fit together with a characteristic structure. Situated environmental fitness often yields a complicated life together. Sometimes this is in symbiosis. Spotted salamander (*Ambystoma maculatum*) eggs are invaded by a green alga that thrives on the nutrients excreted by the developing embryo, the embryo benefiting as the alga removes its wastes and provides oxygen. The tadpoles eat the algae but not before the algae produce motile cells that swim off to invade other egg masses.[9] Sometimes the interdependence is in predation. A herbivore must move to its stationary food, but that requirement alone does not yield much alertness. But since a herbivore is a food for others, and since a carnivore's food moves, the excitement increases by an order of magnitude. Sight, hearing, smelling, speed, and integrative consciousness—all of which contribute to concentrated unity—grow intense just because there is a decentralized interdependence.

The latex in milkweed is toxic to many potential grazers. Monarch caterpillars have evolved a metabolism that tolerates the latex and depend on the toxin to prevent being eaten. Blue jays eat caterpillars that graze on other plants, but not those on *Asclepias*. When caterpillars metamorphose, the toxin is concentrated in wings and legs, so that a jay that grabs a butterfly by the wings will get a bad taste and drop it. Smart jays learn to strip off the wings and legs and eat only the butterfly thoraxes. Juvenile jays do not do this instinctively and are tested for their capacity to learn it.[10] Complexity is an organism-in-environment phenomenon, from the mutations that result in *Asclepias* toxins to the demand for smarter jays. Each kind is molded by the survival pressures of its environment; each kind has to have an adaptive fit. But each kind is creatively pushing to wedge itself in and this pushes the creativity uphill—producing smarter jays, milkweeds with novel chemistries, butterflies with novel metabolisms. And all three—milkweeds, monarch butterflies, and jays—find the system satisfactory enough to flourish in.

An ecosystem has no genome, no brain, no self-identification. It does not defend itself against injury or death as do blue jays and milkweeds. It is not irritable. An oak-hickory forest has no telos, no unified program it is set to execute. But to find such characteristics missing, and then to judge that ecosystems do not count morally, makes another category mistake. To look at one level for what is appropriate at another faults *communities* as though they ought to be organismic *individuals*. One should look for a matrix of interconnections between centers, not for a single center, for creative stimulus and open-ended potential, not for a fixed telos and executive program. Everything will be connected to many other things, sometimes by obligate association, more often by partial and pliable dependencies, and among other things there will be no significant interactions. There will be shunts and crisscrossing pathways, cybernetic subsystems and feedback loops, functions in a communal sense. One looks for selection pressures and adaptive fit, not for irritability or repair of injury; for speciation and life support, not for resisting death.

There is freedom in the interdependencies. The connections are lax; some show up only in statistics. But we do not want to see in the statistics relationships so *casual* that there is no community at all but rather relationships *causal* so as to permit genuine (loose) community, not (tight) organism. Causal links are not less significant because they are probabilistic (as one learns in physics), though they may no longer be determinate. This will be disliked by conservative ecologists and ethicists who, like Einstein, think that dice throwing is irrational. But others find that the looseness is not "noise" in the community; it is a liberal sign of beauty, integrity, and stability.

Not every collection of interacting constituents is a community. Planets form no community; plants do. The latter have an ecology, a home (*oikos*) with its logic (*logos*) of biofunctions and resources. Molecules in a gas and moons around Jupiter have no fitness there, but caterpillars on milkweeds do. The logic of the resident home is as significant as the logic of any ephemeral inhabitant. To praise the individuals (as the creative actors) and to disparage the system (as mere stochastic, inert stage) is to misunderstand the context of creativity.

There is weak holism, not organic holism, though the weakness (if we must use that term) is a strength in the system. The looseness abets community; too much tightness would abort it. It is unlikely that there is any one defining characteristic of or correct approach to such a community. Complex and merging phenomena may be seen from numerous perspectives. Ecology may have to remain (like sociology with its several models of culture) a multiple paradigm science. Its laws will be fewer and more statistical than those in organismic biology, chemistry, or physics, because we are dealing with a community. In another sense, we do not need multiple paradigms *for* the community; *the paradigm is community*. Ecology discovers simultaneously (1) what is taking place in ecosystems and (2) what *biotic community* means as an organizational mode enveloping organisms. Crossing over from science to ethics, we can discover (3) the values in such a community-system and (4) our duties toward it. Interdependence does not always deliver duty, but biological obligation is a relevant consideration in determining moral obligation.

There is a kind of order that arises spontaneously and systematically when many self-concerned units jostle and seek their own programs, each doing their own thing and forced into informed interaction with other units. In culture, the logic of language or the integrated efficiency of the market are examples. No one individual orders either of these, but there is much rationality in both. In nature, an ecosystem systematically generates spontaneous order, an order that exceeds in richness, beauty, integrity, and dynamic stability the order of any of the component parts, an order that feeds (and is fed by) the richness, beauty, and integrity of these component parts. The organismic kind of creativity (regenerating a species, pushing to increase to a world-encompassing maximum) is used to produce, and is checked by, another kind of creativity (speciating that produces new kinds, interlocking kinds with adaptive fit plus individuality and looseness).

When humans evaluate this they have a tendency to think that decentralized order is of low quality because it is uncentered and not purposive; there is no center of experience or control. We do have to be circumspect about "invisible hand" explanations, especially in culture, where there is immorality. In nature there may be clumsy, makeshift solutions. Still, everything is tested for adaptive fitness.

Regularly in ecosystems at least, such order may be a more comprehensive, complex, fertile order just because it integrates (with some looseness) the know-how of many diverse organisms and species; it is not an order built on the achievements of any one kind of thing. A culture is richer, more diverse, more beautiful because it is the product of tens of thousands of minds; it would be quite poor under the centralized control of one mind, or if all thought alike. One mind can provide or appreciate only a fraction of the wealth of a culture. Analogously, ecosystems are in some respects more to be admired than any of their component organisms, because they have generated, continue to support, and integrate tens of thousands of member organisms. The ecosystem is as wonderful as anything it contains. Producing adaptive fits and eliminating misfits, it is the satisfactory matrix, the projective source of all it contains. It takes a great world to breed great lives, great minds.

We cannot admire ecosystems until we see them as places of value capture, which is an aspect of this value integration. One can admire flight in a peregrine falcon, or the gait of a cheetah; but locomotion takes high energy funding. Muscles, nerves, and brains depend, several trophic rungs down the pyramid, on plants (99.9 percent of the biomass) that soak up the sunlight. By their concentration on capturing solar energy, stationary plants make possible the concentrated unity of the zoological world. A result is that animals, the more so the more mobile, select the communities that select them—looseness if you like, but also freedom.

Plants do not intend to help out falcons and cheetahs, nor

does any ecosystemic program direct this coaction. But the system is nevertheless a transformer that interlocks dispersed achievements. Falcons feed on warblers, which feed on insects, which feed on plants; there is a food chain from cheetah through gazelle to Bermuda grass. It is the protein in warblers which falcons can use, that in insects which warblers can use; the energy that plants have fixed is recycled by insects. The kills are the capture of skills. All these metabolisms are as linked as are liver and heart. The equilibrating system is not merely push-pull forces. It is an equilibrating of values.

The system is a game with loaded dice, but the loading is a prolife tendency, not mere stochastic process. Though there is no *Nature* in the singular, the system has a nature, a loading that pluralizes, putting *natures* into diverse kinds, nature $_1$, nature $_2$, nature $_3$ . . . nature $_n$. It does so using random elements (in both organisms and communities), but this is a secret of its fertility, producing steadily intensified interdependencies and options. An ecosystem has no head, but it has a "heading" for species diversification, support, and richness.

We do not want to extrapolate from organism to biotic community, any more than we extrapolate from culture to nature. Rather, we want criteria appropriate to this level. A monocentered organism is a tautology. A monocentered community is a contradiction in terms, though a holistic community is not. Given the logic of ecosystems, there is no reason to shut off value judgments at the skin. We want to love "the land," as Leopold terms it, "the natural processes by which the land and the living things upon it have achieved their characteristic forms (evolution) and by which they maintain their existence" (ecology) (173). The appropriate unit for moral concern is the fundamental unit of development and survival. Loving lions and hating jungles is misplaced affection. An ecologically informed society must love lions-in-jungles, organisms-in-ecosystems, or else fail in vision and courage.

On the scale of decades and centuries, ecosystems undergo succession; on the scale of centuries and millennia, they evolve. Cyclic systems in the short range, they are historical systems in the long range. Fires, floods, disease epidemics, windstorms, volcanic eruptions, and glaciation episodically reset succession, whether in process or at climax. One seldom travels far in a forest without evidence of fire. A majority of the resident species will find niches in the nonclimax stages. On regional scales succession is always somewhere being upset and the nonclimax species migrate accordingly. In result, all phases of succession and associated species are somewhere present. This can seem loose and merely "fortunate." It is also a statistical "law" of ecology, an evolved characteristic of ecosystems, interruptions included, and in that sense not fortunate. As in genetics, only at a different level now, elements of randomness are incorporated in a dynamic life system. A stochastic process is loaded toward richness of life.

In succession, one species pushes out another; the competitions seem noncooperative. But there can be another way of looking at this. The pioneering species gain ground, only to make way for later invaders that replace them. The gray birch succumb to the red maples, and beech later depend on shade provided by the maples. If, in disrupted areas (as all areas eventually are), there are no earlier species that reproduce in the sun, there can be no later species that reproduce only in shade. Water-loving plants invade the margins of a lake; as detritus collects, marsh-loving plants replace them; afterward the bog fills and broad-leaved trees can enter. On Lake Michigan shores, a sand dune starts after an unusually strong wind blowout and thereafter migrates inland. Marram grass can stabilize a dune in a few years, after another decade the grass dies out; jack pine and white pine invade the dune for about a century, and after that black oak replace the pines.

"Each stage reacts upon the habitat in such a way as to produce physical conditions more or less unfavorable to its permanence, but advantageous to invaders of the next stage."[11] Species work themselves out of a home, and leave a place for what comes after. This is not always true; there are perpetually self-regenerating stands; and some species enter communities in spite of, not because of, their predecessors. But one thing that often drives succession is this remarkable competition where winners by their success alter their environment and become losers, a "fortunate" aspect of the "law" of succession, a strange environmental fitness!

Some contend that there is nothing admirable about succession or its periodic upset, nor about beech depending on the maples whose ground they invade. Yet after one has become ecologically sensitive, the system is a kaleidoscope, it turns round with the accidental tumbling of bits and pieces, each with its own flash and color, and yet the whole pattern is also of interdependent parts coacting, patterns repeated over time and topography, endlessly variable, and yet regular, buzzing with life.

And there is much more. Succession is a subroutine in a dramatic story.

It is sometimes said that ecosystems are Markov processes, that is, stochastic systems without long-term memory.[12] The succession from state A (subclimax) to state B (climax) can be specified without attention to history. Whether dice thrown today will come up deuce is independent of what the throws last month were. Whether the maples push out the birch on the east side of the Wisconsin River in the next decade is independent of whether they did this on the west side in the last decade. By contrast, higher organisms and especially persons in cultures are historical entities. Whether the coyote falls for the trap depends on its earlier experiences. Whether a nation passes from state A (peace) to state B (war) depends on lessons learned in the past. As a person matures, the quality of life depends upon cumulative reaction patterns. Ecosystems have no analogous "character." Unlike coyotes or blue jays, they learn nothing. They simply undergo succession, episodically reset. This can seem to give ecosystems less identity and worth in contrast to persons and intelligent organisms. Noncognitive systems deserve little respect, none at all if clumsy and inelegant.

But over evolutionary time ecosystems are quite historical, although decentralized. There is an enormous amount of history in a handful of humus, contrasted with a handful of lunar soil, in the sense that what goes on there bears the memory (cognitive information coded in DNA) of discoveries in previous millennia of Earth's history. This makes biology different from chemistry.

Duties to Ecosystems

One extrapolates mineralogy on Earth to Jupiter and Mars, but one extrapolates nothing extraterrestrially from the birch-maple-beech succession, because earthbound successions are historically evolved phenomena. The birch have "memories," "experiences," strategies accumulated over tens of thousands of years. Black walnuts and *Salvia* shrubs "remember" how to inhibit their competitors. Some of the biochemistries (like photosynthesis) are a billion years old. The behavior of the marmot, hibernating at the onset of winter, is not a probabilistic reaction to cold, but a historically conditioned instinct. An ecosystem has "heritage," a "tradition," the principal cause of what is taking place.

Yet there are surprises. No ecologist can predict successions a thousand years hence, nor are paleobiogeographers surprised when pollen analysis reveals that they were something else a thousand years ago. New historical developments take place.[13] Ecosystems have weak laws and few "constants," only statistical mathematics, no comprehensive theories, all of which can dismay a scientist anxious about predictions. But this can delight the philosopher who finds the laws sufficient constantly to generate history and who finds historical communities satisfying. This liberal environment proves both empirically necessary and sufficient (requisite and satisfying) for producing life, and yet fails to be logically necessary and sufficient (yielding hard deterministic laws). Yet a closed necessary and sufficient environment (a deterministic one) would logically and empirically prevent both the historicity and the individuality we admire within ecosystems. It would block potential and openness. The ecosystem is contingently sufficient for what takes place in it. The ethicist finds this satisfying, contingently sufficient for generating duty.

The species of the community are something like the genes of an organism (said A. G. Tansley).[14] Despite differences between organisms and communities, the analogy correctly teaches that, just as the organism is not its genes but has its history stored there, the community history is not merely that of its species, although the history is written there. The context of history is not all privately in individuals, though the sectors relevant to particular individuals are coded in their DNA strands. History is smeared out across the system. Some is concentrated in the DNA sequences of the birch tree, some in the individual coyote's career, and that diffused in the biotic community as a matrix of coevolved historical centers is equally remarkable. All the transmissible memory is somewhere in the genetic pool, but to think that history is all pinpointed in individuals because the DNA that stores it is within organisms would be as mistaken as to think that human history was all in the books that record it. The impetus for history is as much in the system-place as in the member inhabitants.

A technical way of summarizing this is that communities, not less than individuals, are as idiographic as they are nomothetic.

Perhaps the good of the community, spoken of collectively, is just the goods of individuals distributively, conveniently aggregated, rather like a center of gravity in physics focuses at a point the masses of myriads of particles. Is the "community" a metaphor for the goods of individuals, not something else from it, rather similar to the way in which the goods of United States citizens are summed up as "the nation"? When a hiker has seen all the trees, and asks next, "show me the forest," he has not understood that the forest is nothing more than the trees. Some community-level epiphenomena appear—communities have trophic patterns and organisms do not—but the phenomena are merely interacting individuals. There is complicated life together, but without emergent system-level properties. The system is an ontological fiction. On the other hand, all nominalists soon learn to fear a slippery slope: communities are fictions, their organisms are real; organisms are fictions, their organs are real; organs are fictions, their cells are real; and so on down to quarks.

After one has discarded category mistakes and associated prejudices for skin-in cooperation, centeredness, and so forth, there seems little reason to count one pattern (the organism) as real and another (the ecosystem) as unreal. Any level is real if there is significant downward causation. Thus the atom is real because that pattern shapes the behavior of electrons; the cell because that pattern shapes the behavior of amino acids; the organism because that pattern coordinates the behavior of hearts and lungs; the community because the niche shapes the morphology and behavior of the foxes within it. Being real at the level of community does not require sharp edges, or complex centeredness, much less permanence, only organization that shapes, perhaps freely so, the behavior of member/parts.

Humans may not have duties at every such level of organization. But humans have duties at appropriate survival-unit levels. The organism is one kind of survival unit, as the liver is not. The ecosystem is another critical survival unit, without which organisms cannot survive. The patterns (energy flow, nutrient cycles, succession, historical trends) to which an organism must "tune in" are set "upstairs," though there are feedback and feedforward loops, and system-level patterns are altered by creativity in individual-level mutations and innovations. Editing and support come "from outside" the boundary, "from above" the level of the organism. The community forces are prolific, though they also are stressful forces from the perspective of the individual.

Has the community priority over the individual? Individuals are ephemeral and dispensable, role players in a historical drama where even ecosystems—indispensable and perennial in native-range time frames—enter and exit on geological time scales. The prescription that seems to follow such a description is that communities dominate individuals, since that is the (supposedly admirable) way that nature operates. But moving from *is* to *ought* this way counters the respect for individual autonomy that has become the trademark of liberalism. Community dominance becomes a totalitarian juggernaut. Ethicists would at once censure a social community crushing individuals. Ethics has fought to protect individuals from the tyrannies of culture. Must environmental ethicists reverse hard-won victories and give the community priority? So to trump the individual seems retrogressive.

This fear is a confusion. Had Leopold said, "A thing is right when it tends to preserve the integrity of the human *social* community," he would have on his hands most of the arguments between utilitarian and rights theorists, as well as disputes between liberal and conservative social theorists. A considerable case can be made for the descriptive fact that social forces do shape persons, they induce behavior more than liberals like to admit, and a less considerable case can be made that this ought to be so. Individuals who test whether personal preferences are right by asking what they do to the good of nation, church, or heritage do not always have their priorities reversed. Leopold does think that ethics "tries to integrate the individual to society" and limits freedom in order to favor cooperative social conduct (203). Any social contract theorist would endorse as much.

But Leopold is making no serious claims about interhuman ethics. Nothing really follows from what *is* or *ought to be* in culture to what *is* or *ought to be* in nature. Sociologists, their studies of society in hand, would not tell ecologists what they must find descriptively about ecosystems; that would be a category mistake. Cultures are a radically different organizational mode. Social philosophers, with justice and charity praised in moral society, cannot tell environmental ethicists what is right and wrong in amoral ecosystems, nor what is so when humans deal with ecosystems, radically different from culture.

As humans gain a description of how ecosystems work, Leopold believes that a prescription arises to respect the beauty, integrity, and stability of such systems. That is not all of ethics, only an extension of it. Duties to other humans remain all they have ever been, but "the land" now counts too. Duties to humans (feeding the starving) in conflict with duties to ecosystems (preserving tropical forests) remains a quite unfinished agenda, but Leopold only wanted to start a dialogue that could not begin when the land had no "biotic right" at all (211).

Relations between individual and community have to be analyzed separately in the two communities. To know what a bee is in a beehive is to know what a good (functional) bee is in bee society, but (*pace* sociobiologists) nothing follows about how citizens function in nation-states or how they ought to. And vice versa. So, when humans confront beehives, complaints about a totalitarian society are confusions. Likewise, whether there are duties to ecosystems must be asked without bias from human society. It may be proper to let Montana deer starve during a rough winter, following a bonanza summer when the population has edged over the carrying capacity. It would be monstrous to be so callous about African peoples caught in a drought. Even if their problems are ecologically aggravated, there are cultural dimensions and duties in any solution that are not considerations in deer management.

In biotic communities, the community is the relevant survival unit; its beauty, integrity, and stability come first. Feral goats on San Clemente Island are degrading the ecosystem and authorities are eliminating the goats, overriding individual goat welfare out of respect for the ecosystem. In Yellowstone National Park, when an outbreak of pinkeye threatened the bighorn sheep, park officials refused to treat the disease, although half the herd was blinded and died through starvation and injury. Their argument was that the Yellowstone ecosystem should be preserved as untampered as possible, and that these processes included the struggle between mammals and their natural parasites. Foresters may let wildfires burn, destroying individual plants and animals, because fires rejuvenate the system.

But we need to bring back into focus the looseness, decentralization, and pluralism of biotic communities. Individual organisms are so tightly integrated that we do not term them *communities* at all. No one complains that the goods of heart and liver are only instrumental to the good of the organism. But communities, social or biotic, never have this kind of organization. Biotic communities leave individuals "on their own," autonomous centers, somatic selves defending their life program. (Consistently with this, Yellowstone bighorns should be left "on their own" to combat the *Chlamydia* microbe.)

Ecosystems bind life up into discrete individuals and cast them forth to make a way resourcefully through their environment. So far from being a regimented community, the wilderness has seemed anarchy to many observers, who are more likely to complain of the pulling and hauling that there is no community at all, than to complain that the individual is subordinated to the community. The picture we need, however, is of community that packages everything up into individual lives and binds them together loosely (that is, freely) enough that individuals remain gems in a setting, yet tightly enough that the generating, maintaining system is prior to individual life.

Evolutionary ecosystems maximize individuality in several ways. First, the stochastic contingencies and idiographic historicity that beset every particular organism differentiate their characteristics and fortunes. There is a wildness in ecosystems that resists being completely specified in geology, botany, zoology, and ecology textbooks, even when principles set forth there are coupled with initial conditions. Scientific laws never catch in individual detail all that goes on in a particular place, such as Okefenokee Swamp or Bryce Canyon. Each new lake and canyon will have some differences. No matter how well one knows a particular place, tomorrow and next year will bring surprises.

This is logically and empirically entwined with the heightened individuality of each inhabitant. Each life is given a unique genetic set and lived in a unique place. Some unrelated causal lines and even indeterminate lines meet and make every individual a one-time event. No two coyotes in Bryce Canyon or even two cypress trees in Okefenokee Swamp are alike. Sometimes the differences are insignificant, but sometimes they yield significant individuality. Any organization that removed the diversity, the looseness, the "disorder," the historical particularity of place and individual, would impoverish the ecosystem and the individuality of its member individuals.

Second, evolutionary ecosystems over geological time have steadily increased the numbers of species on Earth from zero to five million or more. Extinction and respeciation have increasingly differentiated natural kinds. Leopold wrote, "Science has given us many doubts, but it has given us at least one certainty: the trend of evolution is to elaborate and diversify the biota" (216). R. H. Whittaker found that on continental scales and for most groups "increase of species diversity. . . . is a self-augmenting evolutionary process without any evident limit." There is a tendency toward "species packing." [15] G. G. Simpson concluded that there is in evolution "a tendency for life to expand, to fill in all available spaces in the livable environments, including those created by the process of that expansion itself. . . . The total number and variety of organisms existing in the world has shown a tendency to increase markedly during the history of life." [16] Islands, though limited in their capacity to carry species, produce isolated kinds. Nature seems to produce as many species as it can; although locally poor in a desert or on a polar ice cap, in the aggregate Earth's ecosystems are many-splendored things. Charles Elton found that five thousand species of animals inhabited two square miles of Wytham Wood in Britain.[17]

Third, superimposed on this increase of quantity, the quality of individual lives in the upper trophic rungs of the ecological pyramid has risen. One-celled organisms evolved into many-celled, highly integrated organisms. Photosynthesis supports locomotion—swimming, walking, running, flight. Stimulus-response mechanisms become complex instinctive acts. Warm-blooded animals follow cold-blooded ones. Neural complexity, conditioned behavior, and learning emerge. Sentience appears—sight, smell, hearing, taste, pleasure, pain. Brains couple with hands. Consciousness and self-consciousness arises. Persons appear with intense concentrated unity, and nature transcends itself in culture. These are liberating developments in the ecosystem; they free individuals. A falcon is more liberated in its ecosystem than is the grass downward in its food chain; it can overlook a territory, migrate, switch prey. This is community looseness now interpreted positively as nourishing individuality.

These developments do not take place in all ecosystems nor at every level. Microbes continue, as do plants and lower animals. These kinds serve continuing roles. All the understories remain occupied. If they did not, the quantity of life and its diverse qualities would diminish. Most creatures are cryptogams, dicots, monocots, fungi, bacteria, protozoans, beetles, mollusks, crustaceans, and the like. Sometimes, there is retrograde evolution, as in tapeworms or viruses, when once free-living organisms lose eyes, legs, metabolisms, or even brains, although retrograde evolution requires that such organisms live in an environment more complex than they are themselves, so that they can borrow their lost skills from their hosts. Meanwhile, the quality of individuality generated at the top rises. So both the quantity and the quality of individuality intensify. This continues despite at least five catastrophic extinctions, so anomalous that many scientists look to extraterrestrial causes: supernovae, collisions with asteroids or oscillations of the solar system above and below the plane of the galaxy. Regardless of their causes, the crashes are followed by swift resurrections, often with novel and advanced forms. Optimization of fitness seems to increase through evolutionary time.[18] There are vastly more individuals and species and they are better fits in their communities.

These developments in natural history are not just a random walk, not just drift. They reveal the rationality of the system, including trial and error. Spasmodic on short ranges, rather like the episodic upset of succession, these prolific trends are a recurrent tendency on long-range scales. Sometimes the evolutionary (as do the ecosystemic) processes seem wandering and wayward, loose, but the results are considerable. There is dice-throwing, but the dice are loaded. The probabilities are showing causal connections, laws.

The community beauty, integrity, and constancy include selecting for individuality. That is a strange, liberating "priority" or "heading" of the system: escalation of individuals in kind and complexity, in quantity and quality, never producing two of a kind exactly alike! That process is as much to be defended as any of its products. The goods and "rights" of individuals (their flourishing and freedom) belong in such a system; the ecosystem itself promotes them in its own way. When humans enter the scene, they should in this respect follow nature. Individual welfare is both promoted by and subordinated to the generating communal forces.

Instrumental value uses something as a means to an end, intrinsic value is value as an end in itself without necessary contributory reference. Leopold laments that nature has previously been considered to have only instrumental value for humans, regarded as the sole holders of intrinsic value. Those sensitive to ecology will revise their axiology. An immediate conclusion is that, apart from any human presence, organisms value other organisms and earth resources instrumentally. Organisms are selective systems. Plants make resourceful use of water and sunshine. Insects value energy fixed by photosynthesis; warblers value insect protein; falcons value warblers. Value capture propels an ecosystem. An organism is an aggrandizing unit on the hunt for instrumental values.

Duties to Ecosystems

Continuing this logic, organisms value these resources instrumentally because they value something intrinsically: their selves, their form of life. No warbler eats insects in order to become food for a falcon; the warbler defends its own life as an end in itself and makes more warblers as she can. A life is defended intrinsically, without further contributory reference—unless to defend the species and that still is to defend a form of life as an end in itself. Such defenses go on before humans are present; and thus both instrumental and intrinsic values are objectively present in ecosystems. The system is a web where loci of intrinsic value are meshed in a network of instrumental value.

Neither of these traditional terms is completely satisfactory at the level of the holistic ecosystem. Member components serve roles, as when warblers regulate insect populations; perhaps that is systemic instrumental value. But, reconsidering, the decentered system, despite its successions and headings, has no integrated program, nothing it is defending, and to say that an ecosystem makes instrumental use of warblers to regulate insect populations seems awkward. We might say that the system itself has intrinsic value; it is, after all, the womb of life. Yet again, the "loose" system, though it has value in itself, does not seem to have any value for itself, as organisms do seem to have.

Nevertheless the system has these characteristics as vital for life as any property contained within particular organisms. Organisms defend only their own selves or kinds, but the system spins a bigger story. Organisms defend their continuing survival; ecosystems promote new arrivals. *Ecosystems are selective systems, as surely as organisms are selective systems.* This extends natural selection theory beyond the merely tautological formulation that the system selects the best adapted to survive. The system selects for what appears over the long ranges, for individuality, for diversification, for sufficient containment, for quantity and quality of life. Appropriately to the community level, ecosystems select employing conflict, decenteredness, probability, succession, and historicity.

We are not any longer confronting instrumental value, as though the system were of value instrumentally as a fountain of life. Nor is the question one of intrinsic value, as though the system defended some unified form of life for itself. We have reached something for which we need a third term: systemic value. This cardinal value, like the history, is not all encapsulated in individuals; it too is smeared out into the system. The value in this system is not just the sum of the part-values. Systemic value is the productive process; its products are intrinsic values woven into instrumental relationships. When humans awaken to their presence in such a biosphere, finding themselves products of this process too—whatever they make of their cultures and an-thropocentric preferences, of their duties to other humans or to individual animals and plants—they owe something to this beauty, integrity, and constancy in the biotic community. Ethics is not complete until extended to the land.

Some will object: cooperation versus conflict, centered cyber-netic autonomy versus loose stochastic process, succession and natural history, systemic values—none of this has touched the nerve of the matter. The final, fundamental problem is that eco-systems have no subjectivity, no felt experiences. Organisms with central nervous systems have psychological life, manifestly present by introspection in human lives, and easy to extend to some nonhumans—more so to chimpanzees, less so to birds. Plants are objects with life, but not subjects of a life. But eco-systems are doubtfully objects at all, rather they are communities mostly of living objects and sparsely of living subjects. No such collection can of itself count morally; any duties must attach to the few subjects who inhabit such places.

An ecosystem cannot be satisfied when given wilderness status; a person can be satisfied when a wilderness is designated. Even coyotes can be satisfied within such a wilderness. Such psycho-logical satisfaction is what the inside/outside issue should have been identifying. The skin is not a morally relevant boundary; rather the boundary is subjective inwardness versus objective me-tabolisms and ecologies. Duties may *concern* ecosystems but must be *to* subjects.

The attractiveness of the duties-to-subjects-only position is that the duties we first know in interhuman ethics are to subjects, and an ethicist can always stipulate that duties stay directed to-ward subjects. A mere object, even one with life, is a misplaced target for duty. This can seem right because it is so familiar. Any duties that involve living or nonliving objects must be reduced to duties to subjects. No doubt some duties do attach only to subjects; some make sense only with human subjects. Persons ought not to be insulted, but squirrels do not suffer much from verbal insults. One may have duties to subjects of a psychological life not to cause needless pain, perhaps not to interfere with their pleasures without justification. We have duties to persons to preserve the integrity, beauty, and stability in the biotic communities that these persons enjoy and resourcefully use. Leopold would endorse such duties, but they are not the only kind he is proposing.

From the ecological point of view the subjectivist position takes a part for the whole. It has a subjective bias. It values a late product of the system, psychological life, and subordinates everything else to this. It mistakes a fruit for the whole plant, the last chapter for the whole story. It orders all duty around an extended pleasure-pain axis, richer through poorer experiences. Such an ethic is really a kind of psychological hedonism, often quite enlightened. But ecosystems are not merely affairs of psychological pains and pleasures. They are life, flourishing in interdependencies pressed for creative evolution. The satisfaction defended at this level is not subjective preferences but the sufficient containment of species.

Leopold does not want a subjective morality but an objective one. He presses this question: Is there any reason for ethical subjects to discount the vital systemic processes unless and until accompanied by sentience? Perhaps to evaluate the entire biological world on the basis of sentience is as much a category mistake as to judge it according to justice and charity found there. The one mistake judges biological places by extension from psychology, the other from culture. What is "right" about the biological world is not just the production of pleasures and positive experiences. What is "right" includes ecosystemic patterns, organisms in their generating, sustaining environments.

True, the highest value attained in the system is lofty individuality with its subjectivity, present in vertebrates, mammals, primates, preeminently in persons. But such products are not the sole loci of value, concentrate value though they may. The objective, systemic process is an overriding value, not because it is indifferent to individuals, but because the process is both prior to and productive of individuality. Subjects count, but they do not count so much that they can degrade or shut down the system, though they count enough to have the right to flourish within the system. Subjective self-satisfactions are, and ought to be, sufficiently contained within the objectively satisfactory system. The system creates life, selects for adaptive fit, constructs increasingly richer life in quantity and quality, supports myriads of species, escalates individuality, autonomy, and even subjectivity, within the limits of decentralized community. If such land is not an admirable, satisfactory, morally considerable biotic community—why not?

Notes

The author gratefully acknowledges critical help from Andrew Brennan and Bruce Omundson.

1. F. E. Clements, *Research Methods in Ecology* (Lincoln, Neb.: University Publishing Co., 1905), 199.

2. H. A. Gleason, "Delving into the History of American Ecology," *Bulletin of the Ecological Society of America* 56, no. 4 (December 1975): 7–10, citation on 10.

3. See an early essay, not published during Leopold's life, "Some Fundamentals of Conservation in the Southwest," *Environmental Ethics* 1 (1979): 131–41. Leopold has abandoned all but the echoes of any philosophy of organism in *A Sand County Almanac*. See J. Baird Callicott, "The Conceptual Foundations of the Land Ethic" (in this volume) for a thorough discussion.

4. See. J. Baird Callicott, "Hume's Is/Ought Dichotomy and the Relation of Ecology to Leopold's Land Ethic," *Environmental Ethics* 4 (1982): 163–74.

5. The best discussion of conceptual issues in ecology is Esa Saarinen, ed., *Conceptual Issues in Ecology*, parts 1 and 2, *Synthese* 43, nos. 1 and 2 (January and February 1980). This collection is largely reprinted as Esa

Duties to Ecosystems

Saarinen, *Conceptual Issues in Ecology* (Dordrecht, Holland: D. Reidel Publishing Co., 1982). For another discussion of "how to think about ecosystems, and how to place them within the scheme of known systems," see J. Engelberg and L. L. Boyarsky, "The Noncybernetic Nature of Ecosystems," *American Naturalist* 114 (1979): 317–24, citation on 323, and in rebuttal Bernard C. Patten and Eugene P. Odum, "The Cybernetic Nature of Ecosystems," *American Naturalist* 118 (1981): 886–95.

6. John Passmore, *Man's Responsibility for Nature* (New York: Charles Scribner's Sons, 1974), 116.

7. D. F. Owen and R. G. Wiegert, "Do Consumers Maximize Plant Fitness?" *Oikos* 27 (1976): 488–92; D. F. Owen, "How Plants May Benefit from the Animals that Eat Them," *Oikos* 35 (1980): 230–35; D. F. Owen and R. G. Wiegert, "Mutualism Between Grasses and Grazers: An Evolutionary Hypothesis," *Oikos* 36 (1981): 376–78, and a subsequent discussion in *Oikos* 38 (1982): 253–59; W. J. Mattson and N. D. Addy, "Phytophagous Insects as Regulators of Forest Primary Production," *Science* 190 (1975): 515–22; S. J. McNaughton, "Serengeti Migratory Wildebeest: Facilitation of Energy Flow by Grazing," *Science* 191 (1976): 92–94; S. J. McNaughton, "Grazing as an Optimization Process: Grass-Ungulate Relationships in the Serengeti," *American Naturalist* 113 (1979): 691–703.

8. This point is elaborated by Paul Shepard, *Thinking Animals* (New York: Viking Press, 1978).

9. P. W. Gilbert, "Observations on the Eggs of *Ambystoma maculatum* with Especial Reference to the Green Algae Found Within the Egg Envelopes," *Ecology* 23 (1942): 215–27, and further comment 25 (1944): 366–69.

10. L. P. Brower and S. C. Glazier, "Localization of Heart Poisons in the Monarch Butterfly," *Science* 188 (1975): 19–25.

11. Clements, *Research Methods in Ecology*, 265. An analogue of this in evolutionary development is the tendency of winners to overspecialize and to become extinct in changing environments, losing out to the less specialized.

12. Henry S. Horn, "Markovian Properties of Forest Succession," in *Ecology and Evolution of Communities,* ed. Martin L. Cody and Jared M. Diamond (Cambridge: Harvard University Press, 1975), 196–211.

13. H. E. Wright, Jr., "The Roles of Pine and Spruce in the Forest History of Minnesota and Adjacent Areas," *Ecology* 49 (1968): 937–55.

14. A. G. Tansley, "The Use and Abuse of Vegetational Concepts and Terms," *Ecology* 16 (1935): 284–307, citation on 290.

15. R. H. Whittaker, "Evolution and Measurement of Species Diversity," *Taxon* 21 (1972): 213–51, citation on 214.

16. G. G. Simpson, *The Meaning of Evolution* (New Haven: Yale University Press, 1964), 243, 341.

17. C. S. Elton, *The Pattern of Animal Communities* (London: Methuen and Co., 1966), 62.

18. D. M. Raup and J. J. Sepkoski, Jr., "Mass Extinctions in the Marine Fossil Record," *Science* 215 (1982): 1501–3.

Appendix

* An Introduction to the
1947 Foreword [to *Great Possessions*]

DENNIS RIBBENS

Great Possessions (Aldo Leopold's own title of what became *A Sand County Almanac*) was submitted to Oxford University Press and to William Sloane Associates with a foreword dated March 4, 1948. This foreword is the one with which readers are now familiar. Another foreword dated July 31, 1947, may be found in the Leopold Papers at the University of Wisconsin–Madison Archives.[1] This was the foreword Leopold had submitted with *Great Possessions* to Alfred A. Knopf September 5, 1947.[2] (Knopf rejected the manuscript exactly one month later.) A note attached to it indicates that Leopold intended to revise the earlier foreword as an appendix for the manuscript which Oxford University Press accepted on April 14, 1948. Because of Leopold's death shortly thereafter, that revision never took place.

Both forewords provide an introduction to the organization and the logic of the book. Both provide an autobiographical framework. But the quantity and emphases of these elements vary greatly between the two. The 1948 foreword, which is much shorter than the 1947 one, elegantly summarizes the logic and organization of the book with relatively little comment about Leopold himself. The 1947 foreword explains the organization of the book only briefly, and for the most part develops the logic of the book less through exposition than through personal reflection. The 1947 foreword makes *A Sand County Almanac* seem much more the evolution and confessions of a conservationist.

APPENDIX

Through the 1947 foreword, *A Sand County Almanac* becomes a
vade mecum on the trail of right thinking about man and land.

The 1947 foreword early on presents Leopold's credentials for
writing about the environment. He cites his experience as a conservation field worker, a research ecologist, a sportsman, and a
naturalist. He claims to approach the issue of land use from these
divers perspectives.

The arguments which follow next in the 1947 foreword most
nearly parallel those of the 1948 foreword. Leopold turns to what
he believed was the very heart of the land use problem—man's
attitude toward land: land should be thought of as more than an
economic resource. Using either similar or identical language,
the two forewords lay out the continuum of man/land attitudes
ranging from land as commodity to land as community, from
man as conqueror to man as fellow citizen, and from human
greed and self-interest to humane love and respect for land.

These ideas were central for Leopold. The most valuable contribution of the 1947 foreword to a reading of *A Sand County
Almanac* lies precisely in the way it aids the reader's understanding of how "these essays describe particular episodes en route" to
Leopold's final "philosophy of land." Fully three-fourths of the
1947 foreword offer specific autobiographical contexts for each of
the essays, from Leopold's experiences as a boy on the banks of
the Mississippi River to his experiences as a mature man on the
banks of the Wisconsin River. The 1947 foreword radiates with
the glow of Leopold's memory of many of his earlier adventures—for example, his trips to the Delta of the Rio Colorado
and to the Sierra Madre areas of Mexico. But it is also remorse-
ful. Leopold candidly confesses his "participation in the extinguishment of the grizzly bear" and admits that he was "accessory
to the extermination of the lobo wolf" during his years as a forest ranger in the Southwest.

The 1947 foreword also contributes to an understanding of *A
Sand County Almanac* by introducing some of the people with
whom Leopold traveled and worked. Those who knew Leopold
well know how important his friends and associates were to the
formation of his environmental thinking, especially after he became Professor of Wildlife Management at the University of
Wisconsin in 1933.[3]

An Introduction to the 1947 Foreword

If "The Land Ethic" is the book's climactic essay, Part I, "A Sand County Almanac," is the section that people most fondly read and return to. While the 1948 foreword provides very little information about the shack background to those essays, the 1947 foreword gives a much simpler, clearer, if more prosaic portrait of the shack, also depicting Leopold as evolving, learning, and changing during his tenure there.

In the course of his autobiographical account, Leopold specifically mentions all of the essays of Part II contained in the manuscript he submitted to Knopf in September 1947, and all of the essays in Part III except "Wildlife in American Culture." Of the essays contained in Part I Leopold specifically mentions only "Great Possessions," singled out for the obvious reason that he had titled the book after it. He generically refers to "a dozen other essays arranged calendar-wise as 'A Sand-Country Almanac.'" Considering how complete Leopold's coverage was for the other sections, this reference probably should also be taken literally as a description of Part I of the September 1947 Knopf manuscript. Part I probably contained, at that time, thirteen essays instead of the present twenty-two. "Good Oak," "The Geese Return," "The Alder Fork," "Prairie Birthday," "The Choral Copse," and "Axe-in-Hand" almost certainly were added in the following months. For Leopold, *A Sand County Almanac* remained an evolving collection of essays. Had Oxford rejected it and had Leopold lived longer, there is no doubt that it would have grown to contain more essays still.

Among the most interesting sections in the 1947 foreword is the last one, in which Leopold provides what he thought to be a proper literary context for his book.[4] It is worth noting that he made a distinction between nature writing before and after access to what he called "our principal instruments for understanding" wild things—ethology and ecology. Good nature writing, he believed, must grow out of acurate observation refined by an ecological view of land.

Although we may never know why Leopold discarded the first foreword in favor of the second, the 1947 foreword can provide readers today with a fresh and informed approach to *A Sand County Almanac*. All of the essays, those of Part II in particular, acquire greater clarity and importance. But most significantly, the 1947 foreword reveals Leopold's own understanding of the odyssey of his life and thought.

Notes

1. Leopold Papers 6B16, University of Wisconsin–Madison Archives.

2. For details of this history, see Dennis Ribbens, "The Making of *A Sand County Almanac*," in this volume.

3. Raymond J. Roark, with whom Leopold traveled to Mexico, was a University of Wisconsin colleague and hunting partner. Franklin Schmidt, Wallace Grange, and Frederick and Frances Hamerstrom were all at one time associated with the central Wisconsin study of prairie chickens. Schmidt, Leopold's first student, died in a fire in 1935. Grange became associated with the Sandhill Game Farm near Babcock. The Hamerstroms, both students of Leopold, continue the study of wildlife near Plainfield, Wisconsin. Hans Albert Hochbaum, also a student of Leopold, was a critical correspondent and early on a collaborator (he was once to have provided the illustrations) with Leopold during the decade-long preparation of *A Sand County Almanac*. See Ribbens, "The Making."

4. Leopold expressly thought himself in the company of Thoreau, Muir, Burroughs, Hudson, Seton, Darling, Lockley, and three contemporary Americans. One can only wonder why Leopold thought his essays had something in common with Sally Carrighar's *One Day on Beetle Rock* (a collection of essays which describe wildlife in the California mountains with an impersonal distance and an absence of either ethical or ecological analysis), Theodora Stanwell-Fletcher's *Driftwood Valley* (primarily a personal account of life in the British Columbia wilderness while doing research on wolves), and *Spring in Washington,* by the literary historian turned naturalist, Louis Halle (which provides a delightful personal description of weather and wildlife, of plants and natural places in Washington, D.C.). Though it is nearer in tone to *A Sand County Almanac* than are the other two, *Spring in Washington* also lacks ecological or ethical commentary. Of the ten authors mentioned, only Muir matches Leopold's brilliant mixture of accurate scientific observation and philosophical analysis.

✴ Foreword

A L D O L E O P O L D

These essays deal with the ethics and esthetics of land.

During my lifetime, more land has been destroyed or damaged than ever before in recorded history. As a field-worker in conservation, I have seen, studied, and measured many samples of this process.

During my lifetime, the stockpile of scientific facts about land has grown from a molehill into a mountain. As a research ecologist, I have contributed to this pile.

During my lifetime, the thing called conservation has grown from a nameless idea into a mighty national movement. As a sportsman and naturalist, I have helped it grow—in size—but so far it has seemed almost to shrink in potency.

This concurrent growth in knowledge of land, good intentions toward land, and abuse of land presents a paradox that baffles me, as it does many another thinking citizen. Science ought to work the other way, but it doesn't. Why?

We regard land as an economic resource, and science as a tool for extracting bigger and better livings from it. Both are obvious facts, but they are not truths, because they tell only half the story.

There is a basic distinction between the fact that land yields us a living, and the inference that it exists for this purpose. The latter is about as true as to infer that I fathered three sons in order to replenish the woodpile.

Science is, or should be, much more than a lever for easier

livings. Scientific discovery is nutriment for our sense of wonder, a much more important matter than thicker steaks or bigger bathtubs.

Art and letters, ethics and religion, law and folklore, still regard the wild things of the land either as enemies, or as food, or as dolls to be kept "for pretty." This view of land is our inheritance from Abraham, whose foothold in the land of milk and honey was still a precarious one, but it is outmoded for us. Our foothold is precarious, not because it may slip, but because we may kill the land before we learn to use it with love and respect. Conservation is a pipe-dream as long as *Homo sapiens* is cast in the role of conqueror, and his land in the role of slave and servant. Conservation becomes possible only when man assumes the role of citizen in a community of which soils and waters, plants and animals are fellow members, each dependent on the others, and each entitled to his place in the sun.

These essays are one man's striving to live by and with, rather than on, the American land.

I do not imply that this philosophy of land was always clear to me. It is rather the end-result of a life-journey, in the course of which I have felt sorrow, anger, puzzlement, or confusion over the inability of conservation to halt the juggernaut of land-abuse. These essays describe particular episodes en route.

My first doubt about man in the role of conqueror arose while I was still in college. I came home one Christmas to find that land promoters, with the help of the Corps of Engineers, had dyked and drained my boyhood hunting grounds on the Mississippi River bottoms. The job was so complete that I could not even trace the outlines of my beloved lakes and sloughs under their new blanket of cornstalks.

I liked corn, but not that much. Perhaps no one but a hunter can understand how intense an affection a boy can feel for a piece of marsh. My home town thought the community enriched by this change, I thought it impoverished. It did not occur to me to express my sense of loss in writing; my old lake had been under corn for forty years before I wrote "Red Legs Kicking." Nor did I, until years later, formulate the generalization that drainage is bad, not in and of itself; but when it becomes so prevalent that a fauna and flora are extinguished.

My first job was as a forest ranger in the White Mountains of Arizona. There I conceived a large enthusiasm for the free life of the cow country, and I admired the mounted cowmen, many of whom were my friends. Through the usual process of hazing and horseplay, I—the tenderfoot—acquired some rudiments of skill as a horseman, packer, and mountaineer.

When the advent of motor transport began to shrink the boundaries of the horse-culture, I realized that something valuable was being lost, but I bowed my head to the inevitability of "progress." Years later, I tried to recapture the flavor of the cow-country in "The White Mountain."

It was in the White Mountain country that I had my first experience with government predator-control. My friends the cowmen shot bears, wolves, mountain lions, and coyotes on sight; in their eyes, the only good predator was a dead one. When some particularly irksome depredation occurred, they organized a punitive expedition, or even hired a professional trapper for a month or two. But the overall outcome was a draw; the predators were kept down, but they were not extinguished. It occurred to no one that the country might eventually become bearless and wolfless. Everyone assumed that the fewer varmints the better, and within limits this was (and is) true.

Then came paid government hunters who worked on salary, took pride in their skill, and (in the case of wolves and grizzlies) were often able to trap a given unit of range to the point of eradication. The sum of a dozen local eradications was extinguishment in the state, and the sum of a dozen "clean" states was national extermination. To be sure, there was a face-saving policy about leaving some predators in the National Parks, but the actual fact is that there are no wolves, and only a precarious remnant of grizzlies, in the Parks today.

In "Escudilla," I relate my own participation in the extinguishment of the grizzly bear from the White Mountain region. At the time I sensed only a vague uneasiness about the ethics of this action. It required the unfolding of official "predator control" through two decades finally to convince me that I had helped to extirpate the grizzly from the Southwest, and thus played the role of accessory in an ecological murder.

Later, when I had become Chief of Operations for the Southwestern National Forests, I was accessory to the extermination of the lobo wolf from Arizona and New Mexico. As a boy, I had read, with intense sympathy, Seton's masterly biography of a lobo wolf, but I nevertheless was able to rationalize the extermination of the wolf by calling it deer management. I had to learn the hard way that excessive multiplication is a far deadlier enemy to deer than any wolf. "Thinking Like a Mountain" tells what I now know (but what most conservationists have still to learn) about deer herds deprived of their natural enemies.

In 1909, when I first moved to the southwest, there had been six blocks of roadless mountain country, each embracing half a million acres or more, in the National Forests of Arizona and New Mexico. By the 1920s new roads had invaded five of them and there was only one left: the headwaters of the Gila River. I helped to organize a national Wilderness Society, and contrived to get the Gila headwaters withdrawn as a wilderness area, to be kept as pack country, free from additional roads, "forever." But the Gila deer herd, by then wolfless and all but lionless, soon multiplied beyond all reason, and by 1924 the deer had so eaten out the range that reduction of the herd was imperative. Here my sin against the wolves caught up with me. The Forest Service, in the name of range conservation, ordered the construction of a new road splitting my wilderness area in two, so that hunters might have access to the top-heavy deer herd. I was helpless, and so was the Wilderness Society. I was hoist of my own petard.

It was at this time that I wrote several papers, now combined in the essay "Wilderness."

Ironically enough, this same sequence of proclaiming a wilder-

1947 Foreword

ness, erasing the predators to increase the game, and then erasing the wilderness to harvest the game, is still being repeated in state after hapless state. The latest instance is the Salmon River, in Idaho.

I have always felt a deep love for canoe trips on wild rivers. In 1922 my brother Carl and I essayed the then wildest stretch of river in the Southwest: the Delta of the Rio Colorado. We were the third party to navigate the Delta, and the first to do it by canoe. Of my many ventures into wild country, this was the richest and most satisfying. I have tried to recapture its flavor, in retrospect, in "The Green Lagoons."

Twenty five years later, while serving on the Wisconsin Conservation Commission, I was impressed by the fact that Wisconsin youth were about to lose one of their last wild rivers: the Flambeau. Most other canoeing rivers in the state had already been harnessed for power. I joined with Conservation Commissioner W. J. P. Aberg and Deputy Director of Conservation Ernest F. Swift in an effort to rebuild a small stretch of cottageless river on the Flambeau State Forest. The defeat of this venture, after it was half completed, by the Wisconsin Legislature, is described in "Flambeau." What is a wild river more or less among farmers thirsty for cheap power?

I moved to Madison, Wisconsin, in 1924, to become Associate Director of the Forest Products Laboratory. I found the industrial *motif* of this otherwise admirable institution so little to my liking that I was moved to set down my naturalistic philosophy in a series of essays: "The Land Ethic," "Conservation Esthetic," and others.

It was at this period that I made a series of vacation trips to the Sierra Madre in Chihuahua, Mexico, in company with my brother Carl, my friend Raymond J. Roark, and my son Starker, by then grown. The Sierra Madre was an almost exact counterpart of my beloved mountains of Arizona and New Mexico, but fear of Indians had kept the Sierra free from ranches and livestock. It was here that I first clearly realized that land is an organism, that all my life I had seen only sick land, whereas here was a biota still in perfect aboriginal health. The term "unspoiled wilderness" took on a new meaning. I recorded these impressions in "Song of the Gavilan" and "Guacamaja."

In 1928 I undertook a game survey for the sporting arms industry, and in 1931 I became Professor of Wildlife Management at the University of Wisconsin. During the ensuing decade several ventures were undertaken which bear upon this autobioraphy.

During the thirties, in company with my friends Franklin Schmidt, Wallace Grange, Frederick Hamerstrom, and Frances Hamerstrom, I did much field work in central Wisconsin. "Marshland Elegy," "The Sand Counties," "Red Lanterns," and "Smoky Gold" express my abiding affection for this region, called poor by those who know no better.

In 1938, with the help of my friend Hans Albert Hochbaum, I helped to organize a waterfowl research station at Delta, Manitoba. I became acquainted with the great marshes of the Canadian wheat belt, and I was shocked to learn how rapidly they were drying up. It was evident that the whole continent was gradually losing its principal nursery for wild fowl. "Clandeboye" is a descriptive sketch of a part of the Delta marsh that seemed to me particularly wild and delightful. I am told that Clandeboye still has water, but it has now acquired roads, empty bottles, and limit-shooting gunners from the States.

One of the penalties of an ecological education is that one lives alone in a world of wounds. Much of the damage inflicted on land is quite invisible to laymen. An ecologist must either harden his shell and make believe that the consequences of science are none of his business, or he must be the doctor who sees the marks of death in a community that believes itself well, and does not want to be told otherwise. One sometimes envies the ignorance of those who rhapsodize about a lovely countryside in process of losing its topsoil, or afflicted with some degenerative disease of its water system, fauna, or flora.

A group of sketches written during the period 1935–1945 deal with this theme of lethal illness, visible only to the ecologist, in the still-lovely landscapes of various states. "Illinois Bus Ride,"

1947 Foreword

"Odyssey," "Cheat Takes Over," and perhaps "On a Monument to the Pigeon" belong in this group. I have been told that "Odyssey" is a complete summary of the fundamentals of ecological conservation.

In 1935 my education in land ecology was deflected by a peculiar and fortunate accident. My family and I had become enthusiastic hunters with the bow and arrow, and we needed a shack as a base-camp from which to hunt deer. To this end I purchased, for a song, an abandoned farm on the Wisconsin River in northern Sauk County, only fifty miles from Madison.

Deer-hunting soon proved to be only a minor circumstance among the delights of a landed estate in a semi-wild region, accessible on week-ends. I now realize that I had always wanted to own land, and to study and enrich its fauna and flora by my own effort. My wife, my three sons, and my two daughters, each in his own individual manner, have discovered deep satisfactions of one sort or another in the husbandry of wild things on our own land. In the winter we band and feed birds and cut firewood, in spring we plant pines and watch the geese go by, in summer we plant and tend wildflowers, in fall we hunt pheasants and (in some years) ducks, and at all seasons we record phenology. All of these ventures are family affairs; to us a landless family, relying on other people's wildlife, has become an anachronism. My experiences at the shack are recorded in "Great Possessions," and a dozen other essays arranged calendar-wise as "A Sand-Country Almanac."

Whatever the philosophical import, or lack of it, in these sketches, it remains a fact that few writers have dealt with the drama of wild things since our principal instruments for understanding them have come into being. Thoreau, Muir, Burroughs, Hudson, and Seton wrote before ecology had a name, before the science of animal behavior had been born, and before the survival of faunas and floras had become a desperate problem. Fraser Darling and R. M. Lockley have expressed, for the British Isles, some fragments of the wildlife drama as illumined by these new viewpoints, but in America, parallel attempts have been few. I salute Sally Carrighar's *Beetle Rock,* Theodora Stanwell-Fletcher's *Driftwood Valley,* and Louis Halle's *Spring in Washington* as among the best of these. My hope is that *Great Possessions* may add something to what they have ably begun.

These essays were written for myself and my close friends, but I suspect that we are not alone in our discontent with the ecological *status quo.* If the reader finds here some echo of his own affections and of his own anxieties, they will have accomplished more than was originally intended.

I take the reader first on a round of the seasons at my shack in Sauk County, Wisconsin, and next on a hop-skip-and-jump tour of the North American continent. In both journeys I sketch the observations and experiences which have impressed me most deeply.

At the end of the volume I try to sum up, in more coherent form, the basic logic of the ecological concept of land.

Aldo Leopold

Madison, Wisconsin
July 31, 1947

Notes

1. The reduction of the Gila deer herd occurred almost a decade before the formation of the Wilderness Society.
2. Leopold became Professor of Game Management in 1933, not 1931.
3. The full title of Sally Carrighar's book is *One Day on Beetle Rock.*

D. R.

Contributors

Index

Contributors

J. BAIRD CALLICOTT, professor of philosophy and natural resources at the University of Wisconsin-Stevens Point, is the coauthor with Thomas W. Overholt of *Clothed-in-Fur and Other Tales: An Introduction to an Ojibwa World View* (Washington, D.C.: University Press of America, 1982), and coeditor with Roger T. Ames of *Environmental Philosophy: The Asian Traditions* (Albany: State University of New York, forthcoming), and author of many articles on environmental ethics.

SUSAN FLADER, professor of American western and environmental history at the University of Missouri-Columbia, has published widely in environmental history, including *Thinking Like a Mountain: Aldo Leopold and the Evolution of an Ecological Attitude toward Deer, Wolves, and Forests* (Columbia: University of Missouri Press, 1974).

PETER A. FRITZELL, professor of English at Lawrence University, has published articles and essays on individual works of nature writing, landscape description, and travel literature in early America, and has contributed to the proceedings of the 1978 National Symposium on Wetlands, *Wetland Functions and Values: The State of Our Understanding*, and to *The Great Lakes Forest: An Environmental and Social History*, edited by Susan Flader (Minneapolis: University of Minnesota Press, 1983).

CURT MEINE is the author of *Aldo Leopold: His Life and Work* to be published in January 1988 by The University of Wisconsin Press.

Contributors

R O D E R I C K N A S H is professor of history and environmental studies at the University of California at Santa Barbara specializing in intellectual and environmental history. His best known book is *Wilderness and the American Mind* (New Haven: Yale University Press, 1967; third edition, 1982).

E D W I N P. P I S T E R is a fishery biologist with the California Department of Fish and Game and is responsible for the biological integrity of around a thousand waters in eastern California. In 1970 he helped form the interagency and interdisciplinary Desert Fishes Council. He has lectured at more than 40 universities in North America and published numerous technical and interdisciplinary articles and essays.

D E N N I S R I B B E N S, University Librarian of Lawrence University, has a continuing interest in Wisconsin writers, and seeks to work out his own relationship with land on his 75-acre farm in Calumet County, Wisconsin. He has both lectured and published articles on Wisconsin writers, nature writers, and on Aldo Leopold.

H O L M E S R O L S T O N, III, professor of philosophy at Colorado State University, has published *Philosophy Gone Wild* (Buffalo: Prometheus Books, 1986), *Science and Religion: A Critical Survey* (New York: Random House, 1987), and *Environmental Ethics* (Philadelphia: Temple University Press, 1987). He is also the associate editor of the journal, *Environmental Ethics*.

W A L L A C E S T E G N E R has taught at the Universities of Utah and Wisconsin, at Harvard, and at Stanford, from which he retired in 1971 as Jackson E. Reynolds Professor of Humanities. He is the author of numerous novels, including *Angle of Repose* (New York: Doubleday, 1971), which won the Pulitzer Prize for fiction in 1972, and *The Spectator Bird* (New York: Doubleday, 1976), winner of the National Book Award in 1977. He is on the Governing Council of the Wilderness Society and has been given the John Muir Award of the Sierra Club. Among his non-fiction works are *Wolf Willow: A History, a Story, and a Memory of the Last Plains*

Contributors

Frontier (New York: Viking, 1962), and *Beyond the Hundredth Meridian* (Boston: Houghton Mifflin, 1962), a biography of Major John Wesley Powell, the first explorer of the Colorado River, and one of the nation's first effective environmentalists.

JOHN TALLMADGE has taught courses and published numerous articles and essays on the literature of natural history at Yale, Dartmouth, and at the University of Utah. For the past seven years he has been a member of the English Department at Carleton College.

BURTON H. CALLICOTT drew the sketches that appear on the part titles. He is professor emeritus at the Memphis College of Arts, and has exhibited his paintings extensively throughout the mid-south. In 1986 he received the Tennessee Governor's Award in the Arts.

Index

A-B cleavage: a pilgrim's progress relating to, 10–11, 221–32; section of "The Land Ethic," 136, 174, 182, 200–201

Acid rain, 239, 242

Aesthetics: Leopold's land/environmental, 9, 157–71; natural, 157–58, 163–69; picturesque, 159–60, 161–62, 169, 170n6. *See also* Beauty; "Conservation Esthetic"

Aiken, William, 206

Alaska Native Lands Act, 240

Albuquerque Chamber of Commerce, 30

"Alder Fork, The" (Leopold), 107, 279

Alfred A. Knopf, 7, 92–94, 96–97, 98–102, 103, 105, 113, 277

Allusion, Leopold's uses of, 117–18, 120

Almanac idea: in Leopold's essays, based on shack experience, 58, 96, 97, 100–101, 103, 104, 287, 288 (*see also titles of essays by name of month*); in other nature writing, 113–14

Altruism, 72, 189, 198

American Association for the Advancement of Science (AAAS), 173

Analogies: in *Almanac*, 124, 131, 132, 153, 161–62. *See also* Metaphors

Animals, liberation or rights of, 6–7, 11, 63–88, 196–98

Anthropocentrism, 66, 69–70

Anthropology, studies of ethics in, 192–93

Antivivisectionism, 68–69, 71

Apache National Forest, 26–27

Archetypal images, in *Almanac*, 124, 134

Arguments (*Almanac's*): respecting the land ethic, 9, 172–85; primary, 130, 137, 139; from implicit to explicit, 135, 137–38; about ecosystems, 137–38, 181. *See also* Ecological arguments

"Arizona and New Mexico" (Leopold), 134

Art appreciation, Leopold analogy to, 161–62

Attfield, Robin, 187, 214–15n4

Audubon Society, 11, 239–40

Austin, Mary, 233

"Ave Maria" (Leopold), 102, 107

"Axe-in-Hand" (Leopold), 102, 279

Bailey, Liberty Hyde, 73

Beauty: Leopold's sense of, 120–21, 158, 160–61, 167, 168, 184; traditional appreciations of, 158–60; human experience of natural landscape's, 160–61; of bogs, 165–66. *See also* Aesthetics (environmental); "Conservation Esthetic"

Bentham, Jeremy, 197

Bergere, Estella, 28–30, 32–33

Index

Bible (environmentalism's), *Almanac* as, 3–4, 233. *See also* Prophet, Leopold as

Biosocial concept, 141; in evolutionary analysis of ethics, 199, 207–8, 211 (*see also* Darwin, Charles); historical anomalies in theoretical foundations of, 209–10

Biosphere (whole), moral consideration of. *See* Ethics; Holism

Biota (the), Leopold on, 262

Biotic community: ecological paradigm of, 139–45, 181–82, 194–96, 198–99, 200, 202; humans as members of, 207, 210–11; questions of respect for nonhuman members of, 208–14 (*see also* Ethics; Rights); Leopold on right and wrong relative to, 228, 246; ecosystem and, 246, 265–68; no shared needs by members of, 252; meaning of, 256. *See also* Ecology; Food chains; Holism; Land pyramid

Biotic rights, of species. *See* Rights

"Biotic View of Land, A" (Leopold), 104, 173–74, 175, 181–82

Birds, Leopold's studies of, 21–25, 36, 59–60

Bogs, beauty of, 165–66

Boone, Daniel, 164

Bradley, Nina Leopold, 12

Brennan, Andrew, 272

British Empiricists, 163–64

Bulletin of the Garden Club of America, 174

Burford, Anne Gorsuch, 239, 240

"Bur Oak" (Leopold), 92, 280n4

Burroughs, John, 112, 116, 287

Callicott, J. Baird, 3–13, 68–69, 75, 80, 82, 85–86n17, 120, 157–71, 186–217

Carlson, Allen, 157n

Carson National Forest, 27–29

Carter, Jimmy, 240

Carver, Jonathan, 106

Category mistakes, 262, 264, 271

Central Wisconsin Conservation Area, 50

Central Wisconsin Foundation, 50

Chamberlain, Neville, 126

"Cheat Takes Over" (Leopold), 287

"Choral Copse, The" (Leopold), 102, 107, 279

Christianity, 176, 199; anthropocentrism of, 70

Civilian Conservation Corps, 34

Civilization, evolution toward, 210–11. *See also* Culture

"Clandeboye" (Leopold), 97, 286

Classics (literary): *Almanac* as, 8, 110–27, 129; other natural history, 110, 113–14 (*see also* *individual authors*)

Clements, Frederic, 200, 246

Climax, of "The Land Ethic," 183–84, 186

Coevolution, 250–51

Cole, Thomas, 112

Colorado River Delta, 31, 118–19

Common Good (the): rights of individuals and, 13n4, 197–98, 202, 205–14, 248–49, 263–72; ecosystems and, 262–68. *See also* Public domain

Community/ies (concept of): land ethic and, 10, 63, 173–75, 177–79, 183–84; intellectual history of, 63–65, 71; in "Thinking Like a Mountain," 126; ethics and, 191–98 *passim*, 200, 205–14; organism analogy abandoned in favor of analogy of, 202; ecosystems and, 255–56; as ideographic and nomothetic, 262. *See also* Biotic community; Ecology; Holism, Humans; Land communities; Land pyramid

Community-individual relations, 13n4, 139–44, 153, 197–98, 202,

Index

205–14, 248–49, 263–72. *See also*
Altruism; Culture; Ethics; Humans; Rights
Competition. *See* Conflicts
Complexity, of ecosystems, 253–55
Compression, concentration (writing techniques), Leopold's uses of, 115–19
Concepts: intellectual history of *Almanac*'s, 6–7, 11, 63–88, 196–97; foundation of land ethic's, 9–10, 34, 186–217; of integrated conservation, 51; key, major, and central, 78–79, 124, 157; formation of, 91–109; biosocial, 141, 199, 207–11; of global village, 192–93, 207; of biological (environmental) integrity, 228–29; in ecology and ecosystems, 272–73*ns*; paradoxical (*see* Paradoxes). *See also specific terms and titles*
Conflicts: of ecological conscience, 128–53; with hydroelectric projects, 227–28; survival and, 249–52. *See also* Community-individual relations
Conscience, environmental, 239–45. *See also* Ecology; Ethics; Right and wrong; Values
Conservation: land ethic, A-B cleavage, and, 10–11, 82–84, 180, 223, 225, 229–31; Leopold and, 21–22, 25, 34–35, 76–81, 221, 281, 282; integrated, concept of, 51; definition of (Leopold's), 57, 221; with liberalism, as environmentalism, 64; role in nature book of issues of, 92–93, 100, 105; ecological conscience and, 174, 179–80, 184; recognized as necessary, 233–45; aesthetics of (*see* Aesthetics; "Conservation Esthetic")
"Conservation Economics" (Leopold), 51
"Conservation Esthetic" (Leopold), 92, 97, 104, 158, 285

"Conservation Ethic, The" (Leopold), 9, 34, 52, 79, 82*n44*, 97; "The Land Ethic" and, 92, 104, 173, 175–76, 178
Conservation movement: *Almanac* and, 37, 81, 83–84, 186; Leopold on, 179–80, 281; in early 1960s, 223–24
Contents (*Almanac*'s): comparison of final with earlier, 102–7; and style in Part II, 130, 133–36, 149–52. *See also specific titles and concepts*
Conversion experience (stories of): Muir's, 112; in Part II, 114; moral of, 125–26
Convictions, Leopold's, 121
Cooper, James Fenimore, 112, 235
Cooperation, in organisms, 247–49, 251
Copernican astronomy, 194, 195
Cranes. *See* "Marshland Elegy"; "Wisconsin"
Creativity, context of, 256–57
Culture: landlessness and, 37; Leopold's essay entitled "Wildlife in American," 97, 104, 279; evolution of, 176; civilization and, 210–11; value of individuality to, 253; ecosystem as analogous to, 257
Curtis, John T., 42

Darwin, Charles, 6, 7; Leopold's reading of, 10, 186, 189, 190; on ethical implications of evolutionary hypothesis, 67–68, 69, 81–82, 189–95, 198–99, 202; Moore's indebtedness to, 70, 71; on altruism, 72, 189, 198; writing style of, 111, 112, 114, 116, 123
Dating (of *Almanac*'s essays), 35–36, 94, 101–7
Death: Leopold's, 60–61; land, 234–35, 286–87
"December" (Leopold), 131, 132–33
Deconstructive thematic critique (of *Almanac*), 8, 128–53

297

Index

Delta, Colorado, 31, 118–19
Delta, Manitoba, 94, 286
Democracy, Golden Rule and, 199.
    See also Freedom
Deontological (is land ethic), 212,
    214
Depression (1930s), 34, 36, 47,
    49–52. *See also* New Deal
Description: essays of, in nature
    book (Leopold on), 91, 92–94,
    98–100, 103. *See also* Field notes
Deserts, fragile ecosystems of, 244
Doctrine, *Almanac*'s, 123–24
"*Draba*" (Leopold), 95, 97, 98, 101,
    104, 107–8, 117, 120
Dualism, of man and nature, 140,
    141–42
Durand, Asher B., 112
Dustbowl, Wisconsin's (1934), 36
Duties, to ecosystems, 11–12,
    246–74

Earth Day, 240
Ecological arguments, Leopold's,
    91–93, 97–98; A. A. Knopf on,
    99–100, 102
Ecological attitude, in writing, 108
Ecological conscience: Leopold's es-
    say entitled, 83, 104, 158; conflicts
    of, 128–53; section in "The Land
    Ethic," 174, 175, 179–80, 183–84,
    212
Ecological essays (Leopold's),
    93–94, 103, 105
Ecological paradigm. *See* Paradigms
Ecological pyramids, 267
Ecological Society of America, 173
Ecology: general awareness of, *Al-
    manac* and, 4, 128–29; intellectual
    history of land ethic and, 6–7, 10,
    81–83, 88*n44*; defined, 63, 179; in-
    troduction of term, 110; general
    principles of, 117; evolution at
    right angles to, 162; on Leopold's
    use of term, 176, 178, 185*ns*; right
    and wrong and, 177; holism of

thought about, 200–202; founder
    of, 246; few paradigms of, 252; as
    multiple paradigm science, 256 (*see
    also* Paradigms); conceptual issues
    in, 272–73*ns*. *See also* Biotic com-
    munity; Community/ies (concept
    of); Individual-community
    relations
Economic determinism, Leopold
    on, 32
Economics: conservation, 51–52; de-
    fined, 63; of land as property, 138,
    237–38, 242–43; land ethic and,
    180, 182, 183–84, 237–38, 243–44
Ecophilosophical questions. *See*
    Conservation ethic; Duties; Eth-
    ics; Questions; Rights; Values
Ecosystems: duties to, 11–12, 246–
    74; arguments about, 137–38, 181;
    explaining, 140–41, 153; models of,
    200; deserts' fragile, 244; com-
    plexity of, 253–55; communities
    and, 255–56; evolutionary, 266–68;
    holism and, systemic value in, 269,
    270; conceptual issues in regard
    to, 273*ns*
Editing, final (*Almanac*), 107–8. *See
    also* Style
Egos (human), ethics and, 253
Eisenhower, Dwight D., 241–42
"Elegy." *See* "Marshland Elegy"
Elton, Charles, 10, 78, 194, 195, 200, 267
Emerson, Ralph Waldo, 112, 117, 237
Empiricists, British eighteenth-cen-
    tury, 163–64
Endangered Species Act (1966), 223
Endeavor Marsh, 36
Energy, ecosystem in terms of, 202–5
Engagement (writing technique of),
    116
Environment: Leopold's legacy and
    the American, 11, 233–45; appre-
    ciating being in naturally beau-
    tiful, 160–61; complexity of parts
    and wholes of, 252–54 (*see also*
    Holism)

Index

Environmental aesthetics. *See* Aesthetics; Beauty; "Conservation Esthetic"

Environmental Defense Fund, 11, 240

Environmental ethics. *See* Conscience; Ethics

Environmentalism: *Almanac* as bible of, 3–4, 233 (*see also* Prophet, Leopold as); Thoreau's moral, 64–65; politics and, 239–44 (*see also* Government; Legislation)

Environmental movement: as inspired by Leopold's philosophy, 229–30; truisms of, 234–35

Erosion, in southwest, 31, 34. *See also* Dustbowl

"Escudilla" (Leopold), 95, 96, 117, 149, 283–84

Essays (Leopold's): kinds and organization of, 91–109; prior to 1941, 92; list of (1944), 96–97; opinions of, 107–8

Esthetics. *See* Aesthetics; "Conservation Esthetic"

Ethical Sequence, The, section in "The Land Ethic," 176–77, 199

Ethics (ecosystems and), 262–72

Ethics (environmental): intellectual history of, 6–7, 12, 63–88, 187–88, 194–200; land aesthetic and, 9, 157–71 (*see also* "Land Ethic, The"); Leopold's learning of, 23; Leopold on changes in, 82–84; questions about, 143, 144 (*see also* Questions); public official's reluctance to deal with, 228. *See also* "Conservation Ethic, The"; Rights

Ethics (general): history of, 188–94, 197–99; anthropological studies of, 192–93; sociobiology and, 213–14; expanding to include land, 238, 270 (*see also* "Land Ethic, The"); egoism in, 253; Leopold on role of, in integrating individual

and society, 264. *See also* Duties; Values

Etter, Alfred, 107

Evans, Edward Payson, 6, 7, 69–70, 72

Evolution: ethical, and land and animal rights, 6–7, 63–88; preservation of species and, 10–11, 204–5; man, land pyramid, community, and, 138, 140–45; ecology at right angles to, 162; cultural, 176; ecological, ethics and, 188–89; of fish and wildlife programs, 226–30; co-, 250–51; Leopold on, 262; and, 266–68; retrograde, 267–68. *See also* Darwin, Charles

Experience (human), as informing thought, 163–64, 166–67; of beauty (*see* Beauty)

Extinctions: rescuing, or preserving, species from, 10–11, 204–5; of fish, 224; individuality, respeciation, and, 267–68

Facts: Leopold's uses of, 116; conflict between values and, 145; illicit derivation of value from, 171*n*13

Fascism, environmental, 13*n*4, 206

Fear, environmental issues and, 242

"February" (Leopold), 132

Feelings. *See* Sentiments

Fertility, land's depleted, 48–49

Field notes, Leopold's, 106. *See also* Description; Naturalist

Fires: peat, 36, 49; repeated slash, 47, 48; near Leopold's shack, 56, 60; Leopold on, 121

Fisheries, management of, 222–30

Fitness, evolution and, 268

Flader, Susan, 4, 5–6, 40–62, 78

"Flambeau, The" (Leopold), 95, 97, 134, 285

Fleming, Donald, 75

Food chains (cycles), 136–38, 205, 258; human-caused breakdowns of, 234–35

Index

Index

Hornaday, William Temple, 29
Humane movement, 68–69
Human rights, ethic of, 193
Humans: relations with nature,
138, 140–53, 208–10; violence of,
in environment, 173–74, 175, 244;
as members of biotic community,
207, 210–11; surviving mechanized, 214; growth of U.S. population of, 238–39; duties to ecosystems of, 263–68, 270–72. *See also*
Community-individual relations;
Culture; Ethics; Man; Rights;
Values
Hume, David, 190, 195, 198, 202,
216*n23*
Hunting: Leopold's and family's,
22–23, 30, 47, 52, 282–83; by
American Indians, 42; A-B cleavage and, 226–27
Husbandry (wild), at Leopold family's shack, 54–61, 287
Hussey, Christopher, 159–60
Hydroelectric projects, conflicts
with, 227–28

"Illinois and Iowa" (Leopold), 134
"Illinois Bus Ride" (Leopold), 95,
119, 286
Impact (influence), *Almanac*'s, 4,
10–12, 74–75, 172, 221–74; essays
on (in *Companion*, Part IV),
221–74. *See also* Conservation
movement; Environmental movement; Ethics; "Land Ethic, The"
Indexing, of "Shack Journal," 106
Indians (Native American): history
of, 42–43; human-nature interaction models from, 208–10
Individual-community relations,
13*n4*, 139–44, 153, 197–98, 202,
205–14, 248–49, 263–72. *See also*
Altruism; Community/ies; Culture; Ethics; Humans
Influence, *Almanac*'s. *See* Impact
Integrity: concept of biological (en-

vironmental), 228–29; of ecosystemic parts, 253
Intellectual history: of Leopold's
ideas, 6–7, 11, 63–88, 196–98;
Leopold's apparent lack of awareness of own, 79–80; of aesthetics,
158–60, 166. *See also* Arguments;
Concepts
Interdependency (interrelatedness):
history of ideas about, 66–67,
85*n13*; lessons in ecological, 118,
194; ethics and, 215*n4*; human-
caused breakdowns in, 234–35;
in ecosystems, 254–58. *See also*
Community/ies; Ecology
Irony, Leopold's, 8, 118–19, 133
Isaac Walton League, 11, 240
Islands, species on, 266–67

"January" (Leopold), 115, 117, 124, 125,
131–32
"January Thaw" (Leopold), 107
Jefferson, Thomas, 46
Journal of Forestry, 173, 174

Kaibab Plateau (Arizona), 34
Kansas plains, Leopold on, 162
Kant, Immanuel, 163–64, 166–67,
190, 197
Keats, John, 127
Knopf. *See* Alfred A. Knopf

Land: Indians', 42–43; white settlement on, 43–47; problems of Wisconsin's sand counties', 47–51;
New Deal and Leopold's working
on problems of, 49–52; Leopold's
central ideas about, 124; as property, or commodity, 138, 237–38;
enlightened decisions about use
of, 158; as surviving mechanized
man, 214
Land aesthetic (Leopold's), ethics
and, 157–58, 163–69. *See also*
Beauty; "Conservation Esthetic"
Land communities: in Part I, 130–

Index

Index

Metamorphosis (animal to, and from, human), in tribal mythologies, 209–10

Metaphors, community as, 262. *See also* Analogies

Misanthropy, ethics and, 71

Mistakes: correcting anthropocentric, 66, 69–70; category, 262, 264, 271

Moore, J. Howard, 6, 69, 70–72

Moral, the (maxim, proverb), in "thinking like a mountain," 125–26

Moral concern, appropriate unit for, 258

Moral environmentalism, Thoreau's, 64–65

Morality: religious tone of Thoreau's, 6, 64–65; from God(s), 189–90; notion of nature's, 210–11. *See also* Christianity; Ethics; Questions; Rights; Values

Morowitz, Harold, 203

Moses, 189, 199

Mouse (field, meadow), 117, 118, 131–32, 152–53

Muir, John, 6, 43–45, 48, 50, 280n4; on respect for, and rights of nature, 66–67, 85n13; writings of, 105, 110, 112; in "Bur Oak," 106; Leopold compared with, 120, 280–81n4

Multiple use, forestry's concept of, 32

Narrative (Leopold's), 114, 115, 119–27, 132–36

Nash, Roderick, 4, 5, 6–7, 11, 63–88

National Environmental Policy Act, 223

National Park Service, 228–29

National Wildlife Federation, 240, 243

Natural aesthetic, 157–71; land aesthetic as new, 168–69. *See also* Beauty; "Conservation Esthetic"

Natural history: *Almanac* and genre of, 8, 110–27, 129; of morals (ethics), 68; Leopold's field notes and, 106; other classics of, 110, 113–14; rationality of developments in, 268. *See also* Description; Nature writing

Naturalist, Leopold as, 21–25, 107

Naturalistic Fallacy, 171n13

Natural resources. *See* Resource management

Natural selection, ecosystems and, 269–70

Nature: humans' relations with, 138, 140–53, 208–10; morality of, 210–11; love of, 235; models of (*see* Ecology; Energy; Organisms)

Nature book, Leopold on kinds of, and organization of, essays for a, 91–109

Nature Conservancy, 11, 240

Nature writing: defined, 111; origins and development of, 111–13; *Almanac*'s differences from earlier, 113–14, 279. *See also* Natural history; Nature book

Necedah National Wildlife Refuge, 50

New Deal: Leopold during, 34, 35–36, 50; projects of, in Wisconsin, 49–52; Leopold's criticism of, 175, 179

New Mexico Game Protective Association (NMGPA), 29–30

Nongame species, increasing emphasis on, 230

Nonhumans (in biotic community), respecting, 208–14. *See also* Animals; Ethics; Species

"Notes for Paper Writing" (Leopold), 104, 105

Noumenal essence, Ouspensky on, 77

Noumenon, Leopold's use of term, 166–68

303

Index

Obligations, mutual, reciprocal, 246–47. *See also* Duties
Observation. *See* Description; Field Notes
Ochsner, Ed, 17, 53
"October" (Leopold), 130–31
"Odyssey" (Leopold), 95, 97, 287
Omundson, Bruce, 272
"On a Monument to the Pigeon" (Leopold), 287
"On Top" (Leopold), 107
"Oregon and Utah" (Leopold), 134
Organism(s): relationships of (pre-ecological models of), 63, 64, 66, 67, 77–78, 200–202; cooperation in, 247–49, 251; ecosystems and Leopold's philosophy of, 251–58, 262–63, 268–69, 270, 272n3; increase in, 266–68
Organization (*Almanac*'s), 4–12, 104–5. *See also* Concepts; Contents
Organizations (environmental), growth of, 239–40
Ouspensky, P. D., 77–78, 79
Outlook, The (section in "The Land Ethic"), 183, 198–99, 204–5, 212
Overpopulation. *See* Population
Owens River, 227, 228
Oxford University Press, 58, 102, 277

Parables (parabolic style), Leopold's, 111, 123–27
Paradigms: ecological, of biotic community, 139–45, 181–82, 194–96, 198–99, 200, 202; ecology's few, and as science of multiple, 252, 256
Paradoxes: *Almanac*'s conceptual, 8; of "The Land Ethic," 142, 211–12; basic, 145; Leopold on, 149; moral, 213–14
Partridge, Ernest, 209–10
Parts I, II, III (*Almanac*'s). *See under titles:* "Sand County Almanac, A"; "Sketches Here and There"; "Upshot, The"

Passmore, John, 186, 246–47
Perception (Leopold on): ecological, aesthetic, 169; developing, 225–26, 236–37. *See also* Aesthetics; Beauty; Ecology; Ethics
Pessimism (Leopold's), about public comprehension and acceptance of land ethic, 83–84
Philosophers (contemporary academic), on "The Land Ethic," 186–87, 201–2, 206–7, 211–12, 214–15n4
Philosophical essays (Leopold's), 92, 97, 101, 102, 103
Philosophical questions, *Almanac*'s grappling with, 8–12, 115, 117–18, 143–46, 148, 151–52
Picturesque, the (in beauty), 159–60, 161–62, 169, 170n6
Pinchot, Gifford, 10, 25–26, 51, 75, 76
Pine Cone (newspaper), 30
Pines: Leopold's planting of, 55–57, 59; Leopold on, 150–51
"Pines above the Snow" (Leopold), 95, 97, 103
Pister, Edwin P., 10–11, 221–32
Plants: consciousness of, 71; rights of (*see* Rights)
Plato, 158, 190
"Plover is Back from Argentine, The" (Leopold), 92
Politics, environmentalism and, 239–44. *See also* Government; Legislation
Population: moral consideration of threatened, 197–98; growth of U.S. human, 238–39; predators and (*see* Predators)
Potter, Van Rensselaer, 75
"Prairie Birthday" (Leopold), 102, 107, 279
Predators, perspectives on, 34, 75, 212, 243, 250–51, 254, 283–85
Preservation: justifications for, 235–36; changed minds about, 243. *See also* Conservation

Index

Price, Uvedale, 159
Pritchard, Rube, 26
Progress: Leopold on, 236, 283; Leopold's ideas as anti-, 237–38
Property, land as commodity or, 138, 237–38, 242–43
Prophet, Leopold as, 186, 214, 233. *See also* Bible (environmentalism's)
Prudential (the), land ethic as, 212, 214
Psychobiotic drama, 133
Publication history (*Almanac's*), 3–4, 58, 74–75, 83–84, 91–109, 113
Public Domain, attitudes toward, 245. *See also* Common Good
Pyramids: of life, 142–43, 144; ecological, 267. *See also* Biotic community; Ecology; Food chains; Holism; Land pyramid

Questions: *Almanac's* raising of, 8–12, 115, 117–18, 143–46, 148, 151–52; about content of a nature book, 91–94, 98–100, 103; of Leopold's attitude, 95–96, 98; of respect for non-humans in biotic community, 208–14 (*see also* Animals); about the "land ethic," 212; about species, 237. *See also* Conscience; Ethics; Rights; Values

Reagan, Ronald, 240, 241, 242
Reason, origin of ethics and, 189–90, 197
Recreation (outdoor): in "Conservation Esthetic," 158; public demands for, 222–23, 224, 237, 244; engineering of, and promoting perception (Leopold on), 225–26, 236–37
"Red Lanterns" (Leopold), 121, 286
"Red Legs Kicking" (Leopold), 282–83
Redwoods National Park, 235
Regan, Tom, 13*n*4, 206
Religion. *See* Christianity; Conver-

sion experience; God(s); Morality; Salvation
Report on a Game Survey of the North Central States (Leopold), 33, 47
Resource management: Pister and, 221–32; practicality of land ethic for contemporary, 230–31; utilitarian philosophy of (*see* Utilitarianism). *See also* Fisheries; Forestry; Game management; Wildlife management
Resurrections, evolutionary, 268
Reverence-for-life (ethic), Schweitzer's, 7, 73–74, 82, 197
Rhetorical figures, Leopold's use of, 118–19
Ribbens, Dennis, 4, 7, 12, 91–109, 113, 124, 277–81
Right and wrong: ecology and, 177; biotic community and, 228, 246; Leopold on, 263–64. *See also* Conscience; Ethics; Values
Rights: animals', plants', rocks', 6–7, 11, 63–88, 196–98; individuals', and good of community, 13*n*4, 197–98, 202, 205–14, 248–49, 263–72; ethic of human, 193; Leopold and (in "The Land Ethic"), 212–14; defining, 215*n*20. *See also* Community-individual relations; Conservation movement; Environmental movement; Ethics; "Land Ethic, The"
Right-to-life, individuals' and community/ies. *See* Individual-community relations
Riley Game Cooperative, 35
Ringland, Arthur, 29
Rivers: Colorado, 31, 118–19; Owens, 227, 228; Leopold on, 285. *See also* "Flambeau, The"
Rolston, Holmes, III, 11–12, 246–74
Romanticism, in nature writing, 112, 113
Roosevelt, Franklin D., 118. *See also* New Deal

Index

Index

237; as like genes, 261; expansion of, 266–68. *See also* Animals; Wolves

Sporting Arms and Ammunitions Manufacturers' Institute (SAAMI), 33, 34, 286

Starker, Charles, 18–21

Stegner, Wallace, 6, 11, 75, 186, 233–45

Steinhacker, Charles, 40*n*

Stochastic: process, ecosystem as, 251; contingencies, 266

Stoddard, Herbert, 33

Strauss, Harold, 92–93

Style (writing), Leopold's, 111–27, 130, 133–36, 139, 141–42, 149–52. *See also specific aspects*

Substitutes for a Land Ethic (Leopold), 180, 196, 212

Succession, in ecosystems, 258–61

Sumner, L. W., 187

Survival, conflict and, 249–51

Symbolic interpretations. *See* Analogies

Sympathy, Darwin's moral, 6, 67–68. *See also* Ethics

Synecdoche, 116–17

Systemic value, 270

Tallmadge, John, 8, 110–27

Tansley, Arthur G., 200, 202, 261

Ten Commandments, 189

Tension zone, Wisconsin's, 42

Text (*Almanac*'s), evolution of, 91–109

Thermodynamic paradigm (of environment), 202

"Thinking Like a Mountain" (Leopold), 27, 38–39*n*12, 77, 95, 96, 98, 105, 284; parabolic style of, 124–27

Thomas, Dylan, 115

Thoreau, Henry David: religious tone of morality of, 6, 64–65; on wildness as world's salvation, 98, 237; writing style of, 105, 112; *Wal-*

den, by, as classic, 110; Leopold's writing compared with, 112, 114, 115, 116, 120, 121, 122; Leopold on, 280*n*4, 287

Title (*Almanac*'s): Leopold's comments about, 98, 105; final decision about, 107. *See also* Great Possessions; *Sand County Almanac, A; and specific essays and parts by title*

Tone (*Almanac*'s), 95–96

"Too Early" (Leopold), 95, 108

Traders, trading (seventeenth- and eighteenth-centuries), 42–43

Turner, Frederick Jackson, 45–47, 49

Udall, Stewart L., 75, 242

U.S. Forest Products Laboratory, 32–33

U.S. Forest Service, Leopold's career in, 22–34 *passim*, 39*n*12, 75, 76, 182, 283–86

Unity: in ecosystems, 252, 253; of individuals in communities, 267

Unity (*Almanac*'s): publication history, and issues of, 95–96, 100–101, 102, 104; moral vision and, 122; parabolic style and, 123–27. *See also* Almanac idea

University of Minnesota Press, 101

University of Wisconsin: Leopold at, 35, 49, 76, 286; Turner at, 47

"Upshot, The" (Part III), 8–10, 103–7, 130, 136–39, 142–46, 151, 157–217

Utilitarianism: in resource management, 10, 221–22; raw motives of (economic determinism), 32; in forest management, 38–39*n*12, 75, 182; in fisheries management, 222–25; of land ethic, 229

Value(s): fundamental, durable, 57, 58; conflict between facts and, 145; illicit derivation from fact(s) to, 171*n*13; moral, 198–99 (*see also* Mo-

Index

DESIGNED BY DAVID FORD
COMPOSED BY G & S TYPESETTERS, INC., AUSTIN, TEXAS
MANUFACTURED BY EDWARDS BROTHERS, INC.
ANN ARBOR, MICHIGAN
TEXT AND DISPLAY LINES ARE SET IN GALLIARD

Library of Congress Cataloging-in-Publication Data
Companion to A sand county almanac.
Includes bibliographies and index.
1. Leopold, Aldo, 1886–1948. Sand county almanac.
I. Callicott, J. Baird.
QH81.L563C66   1987   508.73   87-10396
ISBN 0-299-11230-6

DATE DUE
